TOMATO PLANT CULTURE

In the Field, Greenhouse, and Home Garden

TOMATO PLANT CULTURE

In the Field, Greenhouse, and Home Garden

J. Benton Jones, Jr.

CRC Press
Boca Raton London New York Washington, D.C.

Library of Congress Cataloging-in-Publication Data

Jones, J. Benton, 1930–
Tomato plant culture : in the field, greenhouse, and home garden / by J. Benton Jones, Jr.
p. cm.
Includes bibliographical references (p.) and index.
ISBN 0-8493-2025-9
1. Tomatoes. I. Title.
SB349.J65 1998
635′.642—dc21

98-42149
CIP

Direct all inquiries to CRC Press LLC, 2000 Corporate Blvd., N.W., Boca Raton, Florida 33431.

International Standard Book Number 0-8493-2025-9
Library of Congress Card Number 98-42149
Printed in the United States of America 1 2 3 4 5 6 7 8 9 0
Printed on acid-free paper

Preface

Tomato is the second most commonly grown vegetable crop in the world, potato being number one. Tomato is the number one vegetable grown in home gardens in the United States. Per capita fresh market tomato consumption continues to increase in much of the world, and tomato products are found in a great variety of processed foods. A long-term medical study has revealed that individuals who consume either fresh tomatoes or processed tomato products on a regular basis are less likely to have some forms of cancer than those who do not. Tomato fruit is rich in vitamin A and C and contains an antioxidant, lycopene.

Much information exists on tomato plant physiology and fruit production practices, including environmental and elemental plant requirements. Much of this data, however, is scattered among numerous books, journal articles, and professional and popular press publications. The last major scientific book on tomato, *The Tomato Crop*, was edited by Atherton and Rudich (1986). There have been significant advances made in tomato plant culture, particularly those related to greenhouse production, an industry that is rapidly expanding in many parts of the world. This book consolidates some of the essential data that have been published on tomato culture, focusing on the most recent literature, which includes the cultural characteristics of the plant, fruit production and related quality factors, and the environmental and nutritional requirements for both field- and greenhouse-grown plants.

An interesting book on the history of the tomato in the United States was written by Andrew Smith (1994). The book also includes recipes.

The major objective of this publication is to provide the reader with factual information about tomato plant culture and fruit production, information that will be beneficial to plant scientists and commercial field and greenhouse growers as well as to the home gardener.

About the Author

J. Benton Jones, Jr. is vice president of Micro-Macro International, an analytic laboratory specializing in the assay of soil, plant tissue, water, food, animal feed, and fertilizer. He is also president of his own consulting firm, Benton Laboratories; vice president of a video production company engaged in producing educational videos; and president of a new company, Hydro-Systems, Inc., which manufactures hydroponic growing systems.

Dr. Jones is Professor Emeritus at the University of Georgia. He retired from the university in 1989 after having completed 21 years of service plus 10 years as Professor of Agronomy at the Ohio Agricultural Research and Development Center, Wooster.

He received his B.S. degree from the University of Illinois in 1952 in agricultural science and a M.S. degree in 1956 and a Ph.D. degree in 1959 in agronomy from the Pennsylvania State University.

Dr. Jones is the author of over 200 scientific articles and 15 book chapters, and has written 4 books. He was editor of two international journals, *Communications in Soil Science and Plant Analysis* for 24 years and the *Journal of Plant Nutrition* for 19 years. Dr. Jones is secretary–treasurer of the Soil and Plant Analysis Council, a scientific society that was founded in 1969; and has been active in the Hydroponic Society of America from its inception, serving on its board of directors for 5 years.

He has traveled extensively with consultancies in the Soviet Union, China, Taiwan, South Korea, Saudi Arabia, Egypt, Costa Rica, Cape Verde, India, Hungary, Kuwait, and Indonesia.

Dr. Jones has received many awards and recognition for his service to the science of soil testing and plant analysis. He is a certified soil and plan scientist under the ARPACS program of the American Society of Agronomy, Fellow of the American Association for the Advancement of Science, Fellow of the American Society of Agronomy, and Fellow of the Soil Science Society of America. An award in his honor, the J. Benton Jones, Jr. Award, established in 1989 by the Soil and Plant Analysis Council, has been given to four international soil scientists, one in each of the years 1991, 1993, 1995, and 1997. Dr. Jones received an Honorary Doctor's Degree from the University of Horticulture, Budapest, Hungary; and is a member of three honorary societies, Sigma Xi, Gamma Sigma Delta, and Phi Kappa Phi. He is listed in *Who's Who in America* as well as a number of other similar biographical listings.

Acknowledgments

The author wishes to thank the following individuals for supplying book chapters, research publications, and bulletins which provided information that significantly added to the subject matter presented in this book.

Dr. Stanley Kays
University of Georgia
Athens, GA

Dr. Donald N. Maynard
University of Florida
Bradenton, FL

Mr. Christos C. Mpelkas
Controlled Environment Technologies
Lynn, MA

Dr. A. P. Papadopoulos
Greenhouse and Processing
Crops Research Centre
Harrow, Ontario, Canada

Dr. Mary M. Peet
North Carolina State University
Raleigh, NC

Dr. Richard G. Snyder
Mississippi State University
Crystal Springs, MS

Contents

1 Introduction

CONTENTS

HISTORICAL BACKGROUND

Tomato belongs to the genus *Lycopersicon*, especially *L. esculentum*, that is grown for its edible fruit. The genus *Lycopersicon* of the family Solanaceae is believed to have originated in the coastal strip of western South America, from the equator to about 30° latitude south (Taylor, 1986; Papadopoulos, 1991). The species is native to South America, especially Peru and the Galápagos Islands, being first domesticated in Mexico. In the mid-16th century, the tomato was introduced into Europe, primarily featured in early herbals. It was grown for the beauty of its fruit but was not often eaten, except in Italy and Spain. The fruit was thought to be poisonous like its relative, the deadly nightshade (Heiser, 1969). Although native to the New World, the tomato was introduced back into America from Europe in the 18th century, although its importance as a vegetable has occurred only in this century. It is believed that the American Indians ate the tomato a long time ago.

The botanical classification of the tomato has had an interesting history, first being placed in the genus *Solanum* along with the potato and being identified as *Solanum lycopersicon*. However, this designation was changed to *Lycopersicon esculentum*, *Lycopersicon* being derived from the Greek word meaning "wolf peach," and *esculentum* simply meaning edible. Although there are similar plant characteristics between potato and tomato plants, flower color (yellow for tomato and mostly white or violet for potato) and particularly the shape and manner of the opening of pollen-bearing structures are the characteristics that separate the two plants.

The designation of the tomato fruit as Moor's apple (Italian) or "love apple" (France) during the 16th century is unverified, but commonly believed. The color of the fruit first noted in Italy was yellow. By the 18th century, the tomato began to be used as an edible food, although it was still listed among the poisonous plants.

Although it is certain that the origin of *Lycopersicon esculentum* was South America, the tomato was probably first cultivated and plants selected based on fruit size in Mexico. Therefore, seeds of tomato first taken to Europe came from Mexico after Cortez took Mexico City in 1519, since it was not until 1533 that Peru and Equador were conquered by the Spaniards.

After introduction of the tomato into the United States, it was grown and brought to the table by Thomas Jefferson. George Washington Carver grew and recommended the tomato in an attempt to introduce the fruit into the diet among the poor in Alabama whose diets were woefully deficient in vitamins.

Wild tomato plants are still found from Equador to Chile as well as on the Galápagos Islands, although only two have edible fruit, *Lycopersicon esculentum* (the common tomato in wide cultivation today) and *Lycopersicon pimpinellifolium* (sometimes cultivated under the name of currant tomato). Small fruited type *Lycopersicon esculentum* var. *cerasiforme*, cultivated under the name of cherry tomato, is widely distributed as a wild plant in the tropics and subtropics.

The tomato is an herbaceous perennial, but is usually grown as an annual in temperate regions since it is killed by frost. It originally had an indeterminate plant habit, continuously producing three nodes between each inflorescence, but determinate varieties have been bred with a bushlike form in which the plant is allowed to produce side shoots and the plant terminates with fruit clusters. Determinate varieties have fewer than three nodes between inflorescences with the stem terminating in an inflorescence, producing fruits that are easily machine harvested and primarily grown for processing.

Breeding of the tomato over the past 50 years has substantially changed its plant and fruit characteristics (Stevens and Rick, 1986; Waterman, 1993–1994). Varieties available today for use by both the commercial and home gardener have a wide range of plant characteristics; are resistant to many of the tomato-affected blight and wilt diseases (Stevens and Rick, 1986); and are specifically adapted to a particular set of growing conditions, such as high tropical temperatures (Villareal and Lai, 1978), field and greenhouse conditions, and fresh market versus processing tomato fruit. Maturity dates range from about 60 to more than 95 days, although several 45-day determinate varieties have been introduced for production in the very northern latitudes; and fruit size, color, texture, and acidity can be selected by variety, whether adapted to field or greenhouse conditions, or long or short days. Genetic engineering techniques applied to tomato breeding have been used to produce fruit with a long shelf life (Baisden, 1994). The commercial production of tomatoes in the tropics offers an unique challenge in terms of varieties that can withstand high temperatures, and disease and insect pressures (Cowell, 1979).

Continued interest in the so-called "olde standard garden varieties" (Poncavage, 1997b) or heirloom varieties (Vavrina et al., 1997a; Weaver, 1998) has kept seed supplies still readily available for these older established varieties, while rapid developments in commercial breeding bring a new set of varieties into use yearly as former varieties are discarded (Maynard, 1997). Cluster tomatoes—primarily for greenhouse tomato production, although there is some interest in being field grown (Hochmuth et al., 1997)—are a new increasingly popular type of tomato in which three to four vine-ripened fruits are marketed still attached to the truss stem

FIGURE 1.1 Cluster tomatoes with all fruit attached to the fruiting stem.

(Figure 1.1). Plum or Roma tomato fruits are attracting consumer attention for making sauces and salsa (Benjamin, 1997).

BOTANICAL NAME

Division: Anthophyta
Class: Dicotyledons
Family: Solanaceae
Genus: *Lycopersicon esculentum* Mill — tomato
Lycopersicon pimpinellifolium (L.) Mill — currant tomato
Lycopersicon esculentum var. *cerasiforme* — cherry tomato

COMMON NAMES

The common names for tomato in eight countries are

Country	Name
Danish	Tomat
Dutch	Tomast
French	Tomate
German	Tomate
Italian	Pomodoro
Portuguese	Tomate
Spanish	Tomato
Swedish	Tomat

PRODUCTION STATISTICS

Based on the world Food and Agricultural Organization (FAO) in 1994, tomato fruit—for fresh market and processing—is produced worldwide on approximately 2.8 million hectares (ha) [6.9 million acres (A)] with yearly worldwide fruit production being 77.5 million ton. The top five leading fruit-producing countries are the United States, China, Turkey, Italy, and India. Worldwide tomato production statistics by region of the world and leading countries are given in Table 1.1.

TABLE 1.1
World Production of Tomatoes, 1994

Location	Area (ha × 10^3)	Yield (ton ha^{-1})	Production (ton x 10^3)
World	**2,852**	**27.2**	**77,540**
Africa	428	19.1	8,315
North and Central America	326	45.7	14,874
South America	157	34.0	5,335
Asia	1,313	23.0	30,205
Europe	618	29.7	18,375
Oceania	11	40.3	433
Leading countries			
United States	190	63.7	12,085
China	344[a]	26.0	8,935[a]
Turkey	160[a]	39.4	6,300
Italy	109	48.1	5,259
India	321[a]	15.7	5,029[a]
Egypt	148[a]	31.1	4,600[a]
Spain	62	49.8	3,066
Brazil	58	43.6	2,550
Iran	75[a]	25.9	1,940
Greece	41[a]	44.1	1,810

[a] Estimated.

Source: FAO Production Year Book, 1995, Volume 48, FAO, Rome, Italy.

The demand for the tomato fruit has increased substantially worldwide. Estimated selected vegetable production and daily per capita utilization in developed and developing countries in 1992 based on FAO Production Yearbook (Volume 46, FAO, Rome, Italy) figures were

	Production (10^6 ton)	Utilization (g day^{-1})
Developed countries	33.7	72.7
Developing countries	36.8	23.9

With an increasing world population and efforts to improve the diets of people worldwide, the consumption of vegetables as a portion of the total diet continues to increase. When locally grown fruit is not available, those with high disposal incomes will purchase off-season grown fruits that have been grown either in the field or in the greenhouse at some distance away and shipped to the local market, or grown locally in a greenhouse.

In 1985, per capita consumption of fresh tomato fruit in the United States was 16.6 lb, increasing to 18.8 lb in 1995 (USDA, 1997). It is anticipated that per capita fresh fruit consumption will continue to increase since tomato fruit consumption on a regular basis has been found to have considerable health benefits.

In the United States, Plummer (1992) reviewed tomato statistics from 1960–1990. The most recent statistics can be found in *Agricultural Statistics, 1997* published by the United States Department of Agriculture (USDA, 1997). From that publication, the field production statistics of both fresh market and processing tomatoes for the leading states are given as shown in Table 1.2.

TABLE 1.2
Fresh Market and Processing Tomato Field Production in the United States, 1995

Category	Fresh Market	Processing
Leading states	Florida, California, Georgia	California, Ohio, Indiana
Acres	132,997	330,503
Production	35,025 [1,000 hundredweight (cwt)]	10,831,646 (ton)
Dollar value ($1,000)	996,175	649,610
Fruit yield	263 (cwt/A)	32.73 (ton/A)

Source: U.S. Department of Agriculture (USDA), NASS Vg 1–2, 1996.

Commercial field production of both fresh market and processing tomatoes in the United States (USDA, 1997) in 1985 and 1995 were

				Value	
Year	Area harvested (Acres)	Yield per acre (cwt)	Production (1,000 cwt)	Per cwt (dollars)	Total (1,000 dollars)
		Fresh Market			
1985	129,600	250	32,414	25.90	840,859
1995	132,820	260	30,854	28.50	891,343
		Processing			
		(ton)	(ton)	Per ton (dollars)	Total (1,000 dollars)
1985	257,400	29.56	7,607,690	59.10	449,503
1995	344,380	32.77	11,286,040	63.20	713,544

The field production of fresh market tomatoes has not changed much from 1985 to 1995, while the production of processing tomatoes has increased substantially.

Monthly availability expressed as a percentage of total annual supply for tomatoes was determined by Magoon (1978) based on a total annual production of 2,600 million lb of fruit. For 1995 (USDA, 1997), based on 2,160 million lb of fruit, the flow of tomato fruit into the marketplace has remained fairly constant as is shown in the following comparison:

Month	Percentage of Total Yearly Supply	
	1978	1995
January	7	4.1
February	6	3.8
March	8	3.4
April	9	6.0
May	11	9.9
June	10	11.1
July	11	13.5
August	9	9.1
September	7	11.4
October	8	12.4
November	7	7.8
December	7	7.1

The influx of fresh market tomato fruit from Mexico into the United States is considerable, increasing from 779.5 million lb in 1991 to 1,307.4 million lb in 1995 compared to United States production of 3,388.7 million lb in 1991 to 3,284.0 million lb in 1995 (VanSickle, 1996). The impact of the North American Free Trade Agreement (NAFTA) on Mexican tomato production and importation of fresh fruit into the United States from 1982–1997 has been reviewed by Cantliffe (1997). Greenhouse tomato production is expected to be one of the major developments in Mexico in the near future.

In the United States, the number of tomatoes in foreign trade from 1985 to 1995 has increased substantially, both in terms of imports and exports (USDA, 1997):

Year	Imports		
	Fresh (1,000 lb)	Canned (1,000 lb)	Paste (1,000 lb)
1985	982,270	168,705	111,695
1995	1,702,019	221,894	33,590

Year	Domestic Exports				
	Fresh (1,000 lb)	Canned Whole (1,000 lb)	Catsup/Sauces (1,000 lb)	Paste (1,000 lb)	Juice (1,000 lb)
1985	141,414	10,058	19,472	17,975	1,468
1995	288,021	59,312	252,503	193,215	51,006

OFF-SEASON PRODUCTION

Off-season production of tomato fruit in environmentally controlled greenhouses is rapidly expanding worldwide (Wittwer and Castilla, 1995), particularly in Canada (Carrier, 1997; Mirza and Younus, 1997) and the United States (Snyder, 1993a; Curry, 1997; Naegely, 1997). Greenhouse tomato production up to 1986 has been reviewed by van de Vooren et al. (1986). A global review by Jensen and Malter (1995) discusses the various aspects of protected agricultural potentials, for application in various areas of the world. Janes (1994) has also described tomato production under protected cultivation, while Wittwer (1993) has looked at the worldwide use of plastics for horticultural production. Greenhouse tomato production is big business in western Europe, primarily in The Netherlands (Ammerlaan, 1994), with an expanding acreage in Spain. There is also considerable acreage in England.

Initially greenhouse tomato production was in soil (Brooks, 1969; Wittwer and Honma, 1969), but today much of the current greenhouse tomato production is done hydroponically; and in the future as Jensen (1997) suggests, this method of tomato production is becoming "fashionable again." Greenhouse tomato production is described in detail in Chapter 6.

FIELD- VERSUS GREENHOUSE-GROWN FRUIT

The competition between field- and greenhouse-grown fruit continues although the majority of fresh market fruit is and will continue to be field grown. Today, much of the fruit is being produced some distance from the market. The question of quality between field- and greenhouse-grown fruit is of major importance for the future of the greenhouse tomato industry (Thomas, 1995–1996). In general, greenhouse-grown fruit is vine ripened and can be delivered to the local market within a day or two of harvest. Most field-grown fruit is harvested before the fruit is fully ripe and shipped to the market, ripening occurring either naturally during shipment or by ethylene treatment (Abeles et al., 1992).

Soil field-grown fruit can be coated with soil or dust particles, which although removed by washing prior to placement in the market can affect the self life. Normally the shelf life of greenhouse fruit is better than that of field-grown fruit, which may be due to some soil residue remaining on the fruit. The degree of bruising of the fruit by harvest and handling techniques equally contributes to determining shelf life (Sargent et al., 1997).

Most soil field-grown plants require the use of pesticides and fungicides to keep them pest free, and soils are frequently treated with sterilizing chemicals to eliminate soilborne pests as well as being treated with herbicides to control weeds. Some residues from these applied chemicals can remain on the surface or in the fruit, normally at concentrations well below those considered physiologically significant, and therefore safe for human consumption. However, for some consumers, any presence of applied chemicals on or in the fruit would be considered unacceptable.

Today, an ever increasing quantity of greenhouse-grown fruit is being harvested before being fully ripe and shipped to markets some distance away. For example, greenhouse-grown fruit in The Netherlands is being shipped and marketed in the

United States. Therefore, such sources of fruit will continue as long as shipping costs and fast delivery to the market are economically feasible. Greenhouse production will never be able to match field-grown fruit in terms of volume of supply, but for the quality-demanding customer, greenhouse vine-ripened fruit will remain in demand if of high quality. How this demand is supplied will be determined by a number of factors. If greenhouse tomato plant production can be done pesticide free, the fruit can be so identified in the marketplace, making such fruit of increased value to many consumers.

The *flavor factor* has been (Stevens et al., 1977) and still is the major factor associated with quality (Waterman, 1993–1994; Morgan, 1997; Poncavage, 1997a; Weaver, 1998), particularly for fruit that has been hydroponically grown (Thomas, 1995–1996). Flavor is associated with two factors, genetics and length of time on the vine. Peet (1997) suggests that the excellent flavor and texture consumers associate with homegrown 'beefsteak'-type tomatoes is the standard for comparison with store-available fruit. In today's stores, fruit identified as being vine ripe is frequently the designation used to attract consumers looking for fruit that would have homegrown flavor. Fruit quality is discussed in more detail in Chapter 3.

HOME GARDENING

A 1994–1995 NGA, GIA II survey conducted by the Home Improvement Research Institute found that 78 million Americans garden—42% of all adults—with 18 million new gardeners since 1992, a 30% growth in 3 years. Gardening is the number one leisure activity in America *(American Demographics)*, with baby boomers spending more time in their gardens than in their gyms (*Wall Street Journal*).

The number one garden vegetable is tomato, with some home gardeners just growing a few tomato plants for sufficient fruit to eat or can. The home gardener has several options for growing his or her tomato plants: in a large soil garden, among flowers or other nonvegetable plants, or in some kind of container. Some may even grow them hydroponically. A 1997 *Growing Edge* magazine readership survey revealed that 70% of its readers grow plants hydroponically, and 50% have greenhouses and solariums, with the most popular indoor plants being ornamentals, vegetables, and herbs. This suggests that there are a considerable number of home gardeners growing tomatoes, possibly year round, in an enclosed environment as well as with the hydroponic method.

Most home gardeners tend to select the olde standard varieties (Poncavage, 1997b) and are somewhat reluctant to bring newer varieties into their gardens. Proven performers, such as Better Boy, Whopper, Celebrity, and Mountain Pride, are some of the more commonly grown garden varieties. The so-called heirloom varieties are attracting considerable attention based on the fruit flavor and color (Vavrina et al., 1997a; Weaver, 1998). The only exception would be for the Roma or paste-type tomato for making sauce and salsa, fruit that comes in various shapes (oxhart, plum, pear, long pear, pepper, lemon, and round) and colors (red, red/orange, orange, and yellow). Benjamin (1997) has written about the paste tomato, giving seed sources and varieties suitable for the home garden.

Books that would be useful to the home gardener, providing information on procedures for tomato production, are listed in Appendix I. The *National Gardening Magazine* (Burlington, VT) offers readers valuable information on home garden tomato production, which can be accessed on their web site: www.garden.org/nga/home.html

THE INTERNET

The internet is rapidly becoming a major source of information worldwide on almost any topic, including tomatoes. *A Farmer's Guide to the Internet* has been published by James (1996), providing useful information on how to get started and how to gather information from the internet. Searching the internet for information on hydroponic production, which includes tomato production, has been described by Jones (1996). In the October 1997 issue of the *American Vegetable Grower* magazine, a survey of the top 100 growers found 65% of them using the internet to obtain information and to assist in marketing their crops.

Since the internet is developing so rapidly, it is difficult to advise on the best way to search for information that is available. For the newcomer, it would be best to contact a local provider who can give guidance on the equipment and techniques required for finding specific information web sites.

2 Plant Characteristics and Physiology

CONTENTS

PLANT FORM AND CULTURAL SYSTEMS

PLANT FORMS

Cultivated tomato is divided into two types, *indeterminate* and *determinate*, the former being the single vine type usually trained to maintain a single stem with all the side shoots removed, and the latter terminating in a flower cluster with shoot elongation stopping. Determinate cultivars are usually earlier than indeterminant ones and are especially desirable where the growing season is cool or short, or both. With fruit ripening nearly at one time, it makes this plant type suitable for mechanical harvesting. Indeterminate plants are for long-season production because this form of the tomato plant will continuously produce fruit for an extended period of time if properly maintained. The approximate time from planting to market maturity for an early variety is from 50–65 days while for a late variety from 85–95 days.

FIELD PRODUCTION

For field production, staking the tomato plant results in greater fruit yields than allowing the plant to lie on the ground. The cost for staking and pruning plants to single stem production is a significant factor that must be weighed against potential yield and fruit quality considerations. Field production procedures have been reviewed by Geisenberg and Stewart (1986) and the future of field production, by

Stevens (1986). In the field, keeping the plant free of disease and insect pests is the significant challenge (see Chapter 8). Various plastic culture systems are in common use for fresh market tomato production today (Lamont, 1996). The time period in the field is determined by the length of the growing season. In addition, keeping the tomato plant productive over the whole season may not be possible due to climatic (early or late frosts, drought, excess moisture, etc.) or other conditions (insect and disease pressures).

Greenhouse Production

In the greenhouse, the tomato plant can be maintained for periods of 6 to 9 months in duration, or even longer, by training the plant up a vertical supporting twine, removing older leaves as the lower fruit clusters are harvested, and by lowering the main plant stem to keep the whole plant within easy reach of workers. This process can be sustained as long as the plant is actively growing, free from disease and other stresses. In the greenhouse, it is possible to control the environment and those factors than affect the plant's well-being, and thereby keeping the tomato plant productive over a long period of time (see Chapter 6).

Various systems of plant culture are in use in the greenhouse, or are in the testing and evaluation stage. In one system after the setting of four to five fruit clusters (trusses), the plant is topped to stop any further stem growth. By topping, all the generated photosynthate goes to the fruit already set, which results in large fruit. Fischer et al. (1990), Giacomelli et al. (1993), and Roberts and Specca (1997) have described a single truss system in which the plant is topped after the first fruit cluster is set, and the plant is replaced after the fruit is harvested. Currently, the success of this single-truss system is based on the ability of each plant to produce at least 2 lb of fruit. This system of fruit production allows for a unique design in terms of method of growing (ebb-and-flow hydroponics) on moving trays of plants that utilizes all the growing space in the greenhouse, and uses an automated system that brings the growing trays to the workers.

Normally, the first two to three fruit clusters contribute more than 50% of the fruit yield obtained in the first four to six clusters, and this rhythm of fruit production is sustained with the continuing growth of the plant. As fruit is removed from the initial (lower) clusters, additional fruit is set on the developing clusters as the plant tends to maintain a balance of fruit based on its leaf area and growing conditions, particularly that of light (both intensity and duration). Successful fruit production in the greenhouse has been described by some as those conditions and procedures that will sustain fruit production at levels between 2.0 and 2.5 lb of fruit per plant every 7 days over an extended period of time.

FLOWER CHARACTERISTICS

The tomato plant is day neutral, flowering under conditions of either short or long days; therefore, it is widely adapted for production at most latitudes.

The inflorescence is a monochasial cyme of 4 to 12 perfect and hypogynous flowers. Primitive tomatoes have the solanaceous trait of five flower parts, but modern tomato varieties often have more than five yellow petals and green sepals. The five anthers are joined around the pistil in *Lycopersicon*, one of the key distinctions from the closely related *Solanum* genus. Wild *Lycopersicon* species, which are self-incompatible and therefore are obligate cross pollinators, have their style exserted beyond the anther cone. Cultivated tomatoes are self-fertile, and their style length is similar to the anther length, a characteristic that favors self-pollination.

Although flowers will self-pollinate, physical vibration of the flower either by mechanical means or by insects is essential for complete pollination to produce fruit normally shaped and symmetrical. Flower abortion occurs when low light conditions exist and when the plant is under stress. Flowers that are not pollinated will abort.

There is a significant positive relationship between mean daily radiant exposure (400–700 nm) and number of flowers reaching anthesis in the first inflorescence, the maximum number of flowers occurring at approximately 1.0 megajoule (MJ) m^{-2} day^{-1} (Atherton and Harris, 1986). Plant density is also a factor that can influence flower abortion and development; loss of flowers occurs with increasing density from no flower loss at 5 plants per square meter to 90% loss with 30 plants per square meter. These same authors have described the other important environmental and cultural factors that affect the setting of flowers and floral development.

POLLINATION

Current commercial tomato varieties are self-pollinating, and bees are not normally needed unless the air is still and the air temperature is cool. Optimum nighttime temperature for pollination is between 68 and 75.2°F (20 and 24°C) (Peet and Bartholomew, 1996). These same authors found that with less than an 8-h photoperiod, or low irradiance, or both, all flowers aborted at 86°F (30°C) nighttime temperature. Kinet (1977) found that by doubling either the photoperiod at low irradiance or the irradiance at the same photoperiod, abortion was significantly reduced.

Pollination will occur when the nighttime temperature is between 55 and 75°F (13 and 24°C) and when the daytime temperature is between 60 and 90°F (15.5 and 32°C). At higher or lower temperatures, particularly at night, flowers will drop without setting fruit.

When hand pollinating using a vibrator (Figure 2.1), the open flower must be vibrated several times over several days to ensure complete pollination. The vibrator probe is placed on the underside of the truss stem next to the main stem as pictured in Figure 2.2. If pollen is ready to be released, a small cloud of yellow pollen will be seen falling from the open flower when vibrated. Great care needs to be taken to keep the vibrating probe from hitting the flower or any small developing fruit because contact will scar the fruit. Bumblebees, when placed in the greenhouse to pollinate the flowers, can damage flowers (they may abort) if there are insufficient flowers to pollinate. If a tomato flower has been visited by a bumblebee as shown in Figure 2.3, the tip of the yellow flower around the stigma will darken as shown in Figure 2.4.

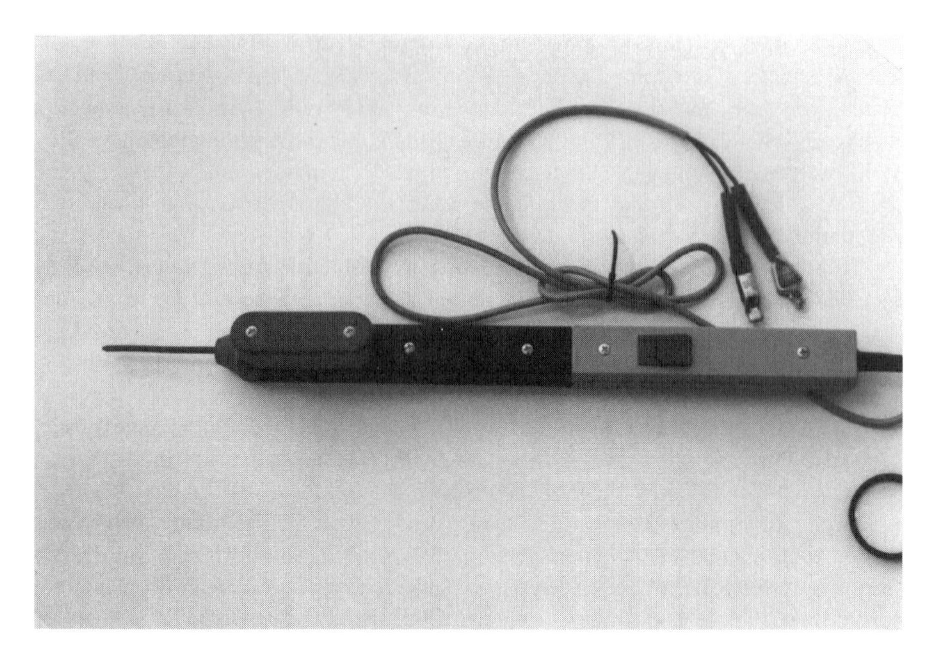

FIGURE 2.1 Hand vibrator for pollinating tomato flowers.

FIGURE 2.2 Mechanical vibrator (coming from the left) probe properly placed below the flower stem.

FIGURE 2.3 A bumblebee visiting a tomato flower.

FIGURE 2.4 The scarring of a tomato flower showing that the flower has been visited by a bumblebee.

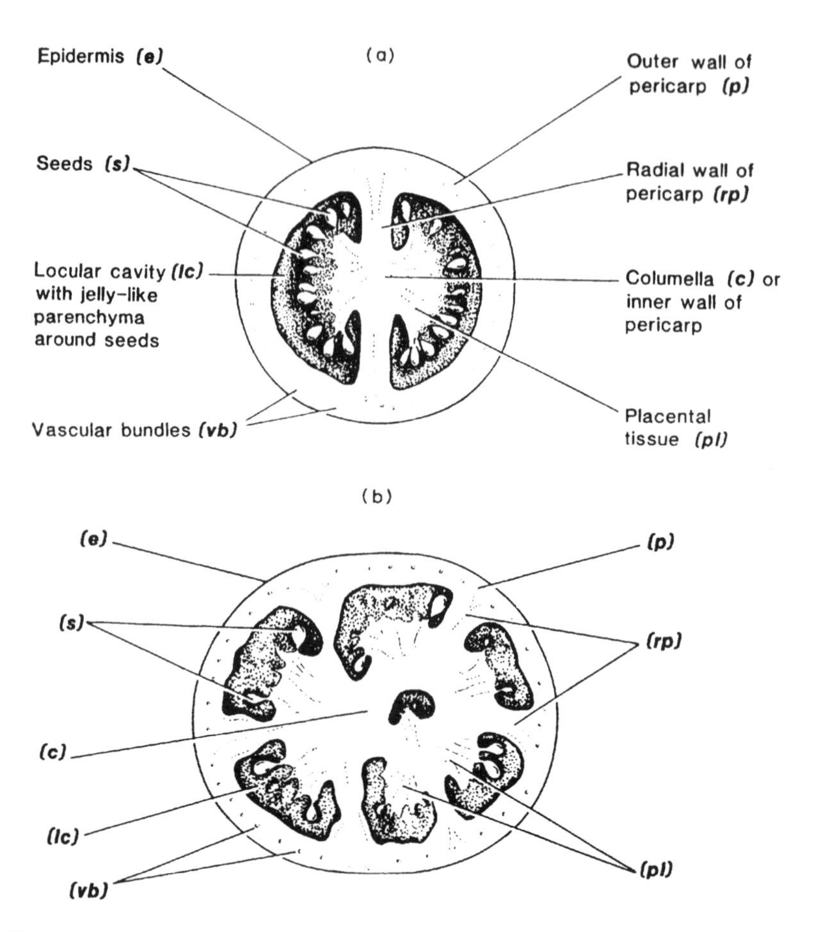

FIGURE 2.5 The anatomy of a two-vascular (top) and multivascular (bottom) tomato fruit. (From Ho, L.C. and J.D. Hewitt. 1986. p.123 In: Atherton, J.G. and J. Rudich (Eds.), *The Tomato Crop: A Scientific Basis for Improvement*. Chapman & Hall, New York. With permission.)

FRUIT CHARACTERISTICS

The tomato fruit is a berry with 2 to 12 locules containing many seeds (Figure 2.5). The size and shape of the fruit itself is affected by the extent of pollination, which in turn determines the number of seeds filling each locule. There is a substantial range in fruit characteristics among the many tomato varieties available today. Vast improvements have been made through breeding with the development of hybrids that are disease (primarily virus) resistant, and have improved yield potential and a range in fruit quality characteristics (fruit shape, color, acidity, etc.). The most recent development has occurred with varieties that have a long shelf life (Thomas, 1995–1996).

Most tomato varieties are red in color due to the red carotenoid lycopene. Different single genes are known to produce various shades of yellow, orange, or green fruit. The yellow color is not related to the acidity of the fruit. Pink fruit is due to a single gene (Y) that prevents formation of the yellow pigment in the epidermis of the fruit.

More than 90% of the fresh weight of the tomato fruit is water, and the availability of water to the plant can influence fruit size. As the tomato fruit develops, the percentage of fresh weight that is sucrose decreases while starch and reducing sugars increase. The composition of tomato fruit is discussed in more detail in Chapter 3 of this book.

Fruit size is strongly influenced by solar radiation, decreasing with decreasing intensity, and by plant spacing, decreasing with increasing plant population (Papadopoulos and Pararajasingham, 1996). Fruit number per truss was found to be positively correlated with solar radiation received, particularly when less than 1.5 MJ m^{-2} day^{-1} (Cockshull et al., 1992). The effect of different light intensities (1.01–3.46 MJ m^{-2}) and the corresponding CO_2 content (350–1000 vpm) of the air surrounding the tomato plant affects the rate of photosynthate transport and photosynthesis in the leaves (Ho and Hewitt, 1986). Similar rates were observed for the combination of 1.01 MJ m^{-2} of radiation with 1000 vpm CO_2 and 1.30 MJ m^{-2} of radiation with 350 vpm CO_2. The overall impact of radiation, however, was more significant than CO_2 content. However, increasing the CO_2 content of the air can partially compensate for the effect of low light intensities on photosynthate production.

Removal of fruit tends to increase the size of remaining fruit, and removal of fruit from lower trusses increases fruit size on remaining trusses. Topping the plant also results in increased fruit size. Even though the harvest index (fruit biomass per total biomass yield) for tomato is on the order of 65–72%, fruit yield is probably source limited. Fruit size and number and total yield are related because total yield tends to remain constant under a set of environmental conditions, while size and number of the fruit will adjust to equal the total yield potential (i.e., large fruit fewer in number or smaller fruit larger in number).

Fruit yields are variously expressed, which makes comparisons among varieties and growing conditions difficult to make. In general, fruit yields are expressed as total weight of fruit for the season. Recorded fruit yields based on per plant production over a specified time period are becoming more common. Recording fruit yields per unit of occupied space is another method of yield expression although total weight of yield is frequently not significantly affected by the number of plants per unit of space but fruit size is. With decreasing fruit weight, there is usually an increase in fruit numbers (smaller fruit).

Expressing fruit yield on the basis of a time factor is what is needed, a useful comparative number being the fruit yield per plant per 7 days of production. Many greenhouse growers have set their production goal at 2.0–2.5 lb of fruit per plant every 7 days when the plant is in full production. The ability to achieve that goal and sustain it over an extended period of time is the challenge. However, the production capability of any system must measure yield over an extended period of time greater than 7 days. Therefore, a standard method of fruit yield determination is needed.

The genetic yield potential for the tomato plant is not known, although the ability of growers to obtain increasingly higher yields over the past decade suggests that the genetic potential has not been achieved with the commonly grown varieties. It would be interesting to know what the genetic potential is, and then to establish those environmental factors that would be required to achieve that potential.

TEMPERATURE REQUIREMENTS

The tomato plant prefers warm weather because air temperatures, 50°F (10°C) or below, will delay seed germination, inhibit vegetative development, reduce fruit set, and impair fruit ripening. The tomato plant cannot tolerate frost. High air temperature, above 95°F (35°C), reduces fruit set and inhibits development of normal fruit color. The optimum range in air temperature best suited for normal plant growth and development and fruit set is between 65 (18.5°C) and 80°F (26.5°C), with day and nighttime temperature ranges being 70–85°F (21–29.5°C) and 65–70°F (18.5–21°C), respectively. The growing degree day base temperature is 51°F (10.5°C), a temperature below which growth is negligible; while on the contrary, the best growing temperature is 70–75°F (21–24°C), the minimum being 65°F (18.5°C) and the maximum being 80°F (26.5°C).

Although air temperature is critical for normal vigorous plant growth, the canopy (leaf) temperature may be far more important, a factor that can be controlled in environmental chambers and greenhouses, the optimum range being between 68 and 71.6°F (20 and 22°C). The combination of air temperature, relative humidity, and plant transpiration rate will determine the canopy temperature.

A tomato plant exposed to cool air temperature, less than 60°F (15.5°C) for extended periods of time, will begin to flower profusely with flower clusters appearing at terminals, typical of determinate plants. Two flowers may fuse together forming an unusually large flower. Flowers will remain open on the plant for several weeks without the formation of fruit. If a fruit does appear, it may be ribbed in appearance indicating incomplete pollination, or two or more fruit may fuse together.

Air temperature can have a marked affect on the atmospheric demand (moisture requirement) of the tomato plant, increasing with increasing air temperature. For example, Geisenberg and Stewart (1986) gave the water requirement for field tomatoes as 2,000–6,600 m^3 ha^{-1} under normal air temperatures, but the range was from 8,000–10,000 m^3 ha^{-1} for extremely warm desert conditions. A mature tomato plant may wilt during an extended period of high air temperature if the plant is not able to draw sufficient water through its roots, a condition that can occur if the rooting medium is cool or the rooting zone is partially anaerobic. Also the size of the root system may be a factor. Just how large the root system must be to ensure sufficient rooting surface for water absorption is not known. However, the relationship between air temperature and relative humidity can moderate the transpiration rate, reducing the atmospheric demand with increasing humidity.

PHOTOSYNTHETIC CHARACTERISTICS

When chlorophyll-containing plant tissue is in the presence of light, three of the essential elements, carbon (C), hydrogen (H), and oxygen (O), are combined in the process called *photosynthesis* to form a carbohydrate as is illustrated in the following:

carbon dioxide ($6CO_2$) + water ($6H_2O$)
(in the presence of light and chlorophyll)

↓

carbohydrate ($C_6H_{12}O_6$) + oxygen ($6O_2$)

Carbon dioxide (CO_2) is from the air, and water (H_2O) is taken up through the roots. A water molecule is split and combined with CO_2 to form a carbohydrate while a molecule of oxygen (O_2) is released. Tomato is a C3 plant since the first product of photosynthesis is a 3-carbon carbohydrate.

The photosynthetic process occurs primarily in green leaves and not in the other green portions (petioles and stems) of the plant. The rate of photosynthesis is affected by factors external to the plant, such as:

- Air temperature (high and low)
- Level of CO_2 in the air around the plant
- Light intensity and quality

Under most conditions, both in the greenhouse as well as outdoors, the energy level impacting the plant canopy is the factor that influences plant growth; and for tomato, that energy level is frequently exceeded. In any growing system, the ability to control both the total amount of energy received over a period of time and the energy level at any one point in time determines plant performance (Mpelkas, 1989).

The photosynthetic saturation point for tomato lies between 600 and 800 μmol m^{-2} sec^{-1}, not too dissimilar to other C3-type plants whose first product of photosynthesis is a 3-carbon sugar phosphate. The photosynthetic rate is linear for C3 plants from 250–900 MJ m^{-2}, the light conversion coefficient being about 1.8 g dry weight per megajoule total solar radiation. The tomato plant can grow well in continuous light between 400 and 500 μmol m^{-2} sec^{-1}. Manrique (1993) reported that the tomato light saturated at 13 MJ m^{-2} day^{-1}.

Papadopoulos and Pararajasingham (1996) observed light saturation of tomato leaves inside the canopy at 170 W m^{-2}, and 210 W m^{-2} for outer leaves. Measurement of solar radiation in the greenhouse at midday was 400 W m^{-2} in June with the range in mean solar radiation being from a high of 730 W m^{-2} (June) to a low of 220 W m^{-2} (December). The lower measurements inside the greenhouse reflect the effect of structural features that block incoming radiation.

The integrated photon flux, CO_2 air content, and atmospheric humidity are the critical parameters, a photon flux of 20–30 μmol m^{-2} day^{-1} being optimum for most plants, including tomato. The photon flux measured at a point in time times 0.0036 will give moles per square meter per day.

Photosynthetic active radiation (PAR) is that portion of the light spectrum (400–700 nm) that relates to plant growth; a brief review on this subject has been given by Davis (1996). Light measurements for photosynthesis are normally expressed as photosynthetic photon flux density (PPFD).

Fruit production is directly related to solar radiation as Cockshull et al. (1992) found in the United Kingdom; yields of 2.01 kg of fresh weight were harvested for every 100 MJ of solar radiation from February to May, and during the longer light days the fruit yields were 2.65 kg for the same light energy input. In The Netherlands, De Koning (1989) reported 2.07 kg of fresh fruit produced per 100 MJ of solar radiation. The two major limiting factors in the greenhouse tomato production in the northern latitudes is low light intensity and short day lengths.

Light interception by the plant canopy is influenced by the leaf area exposed to incoming radiation with plant spacing having a significant effect on interception. Papadopoulos and Pararajasingham (1996) have studied plant spacing effects on crop photosynthesis. One of the primary reasons why greenhouse tomato yields far exceed that obtainable for field-grown plants is the greater interception of light energy due to the increased leaf area indices of the greenhouse plants. The value of the lower leaves on the tomato plant is considerable in terms of their contribution to plant growth and fruit yield.

In layman's terms, one could correlate light intensity, when less than the saturation level, to per plant fruit production during sustained fruit set as:

Light conditions	Pounds fruit per plant per week
Very low	0.5
Low	0.6–1.0
Moderate	1.1–1.5
Good	1.6–2.0
High (at saturation)	2.1–2.5

High light intensity is probably as detrimental to tomato fruit production as low light intensity is. With high solar radiation impacting fruit, cracking, sunscald, and green shoulders can be the result. In addition, high light intensity can raise the canopy temperature, resulting in poor plant performance. In southern latitudes and during the summer months in all latitudes, greenhouse shading is essential to maintain production of high quality fruit. The author has noted that tomato plants grown outside in Georgia (U.S.) during the summer months grow and produce better when in partial shade (particularly at solar noon) or when under some kind of over-the-top plastic cover that partially shades the plant.

Under low light conditions, light supplementation is more effective by extending the hours of light rather than attempting to increase light intensity during the sunlight hours. Supplemental light at 100 µmol m^{-2} for an 18-h day was found to increase fruit yield 1.8-fold (Manrique, 1993).

Based on scientific terms for the production of greenhouse tomatoes, control of the light and air environment could be described as "process management" of the growing system in which assimilation, translocation, allocation, and uptake are the factors requiring control to maintain a vigorously growing tomato plant and high fruit yields; these factors are becoming controllable as greenhouses install computer-directed control devices.

Plant factors that can impact photosynthetic efficiency are the water status of the plant (level of turgidity) and nutritional condition. Loss of turgidity will significantly reduce the photosynthetic rate due in part to the closing of leaf stomata. Several of the essential plant nutrient elements—particularly copper (Cu), manganese

(Mn), iron (Fe), zinc (Zn), phosphorus (P), and magnesium (Mg)—are elements directly involved in energy transfer reactions (see Table 4.2), reactions that directly impact photosynthesis (Porter and Lawlor, 1991). Potassium is involved in the opening and closing of stomata; its inadequacy would then impact the photosynthetic rate.

LIGHT QUALITY

Plants respond to both light intensity and quality. When there is excess blue light with very little red light, the growth will be shortened, hard, and dark in color; if there is excess red light over blue light, the growth will become soft with internodes long, resulting in lanky plants. The author saw these effects in two greenhouses located close to each other, one glass covered and the other covered with fiberglass. The tomato plants in the glass-covered house were tall and light green in color, while those in the fiberglass-covered house were short and dark green in color; the differences in plant appearance were due in part to wavelength light filtering. However, fruit yields and quality were comparable in both houses; although in the fiberglass-covered house, the cultural requirements were easier to manage with shorter plants. The growth response of the tomato plant to light quality is shown in Table 2.1

TABLE 2.1
Tomato Response to Quality of Light

	Type of Light					
Plant Response	**Blue**	**Green**	**Red**	**Infrared**	**White**	
Height of plants (cm)	29.3	30.6	31.6	41.4	20.0	2.51[a]

[a] Significant difference at $p = 0.05$.

Source: Wang, 1963.

There seems to be some evidence that diffuse light may also be a factor affecting plant growth, a factor that has not be adequately explored.

CARBON DIOXIDE

The normal atmosphere contains about 300 mg L^{-1} [parts per million (ppm)] carbon dioxide (CO_2); and in a tomato greenhouse canopy, it can be quickly drawn down to 200 mg L^{-1} (ppm) (Bruce et al., 1980). Carbon dioxide level is thought not to be a problem if the normal atmospheric level can be maintained in the plant canopy. However, the tomato plant, being a C3 plant, is highly responsive to elevated CO_2 in the air surrounding the plant. In a greenhouse, elevating the CO_2 content to 1000 mg L^{-1} (ppm) can have a significant effect on the tomato plant growth and yield (Ho and Hewitt, 1986). Net photosynthesis for various leaves in the plant canopy will range from a low of 56 μg CO_2 m^{-2} sec^{-1} to a high of 80 μg CO_2 m^{-2} sec^{-1} (Papadopoulos and Pararajasingham, 1996).

Under high light intensity with 1000–1500 mg L^{-1} (ppm) CO_2 levels, tomato leaves became thickened, twisted, and purple; and the intensity of deformity increased with increasing CO_2 concentration. Schwarz (1997) has described the conditions and symptoms of CO_2 toxicity that normally occurs when the CO_2 content of the air is greater than 1000 mg L^{-1} (ppm).

WATER REQUIREMENT

The tomato plant needs plenty of water but not an excess because tomato roots will not function under waterlogging (anaerobic) conditions. When the moisture level surrounding the roots is too high, epinasty, poor growth, later flowering, fewer flowers, and lower fruit set occurs; and fruit disorders such a fruit cracking will occur when water availability is inconsistent. Rudich and Luchinisky (1986) have reviewed the water economy of the tomato plant in relation to its growth and response to varying water levels. The tomato plant responds quickly to fluctuations in radiation, humidity, and temperature, factors that significantly impact the plant. The ability of the tomato plant to adjust to these conditions determines the rate of plant growth as well as the yield and quality of fruit. Even under moderate water stress, photosynthesis is slowed because the movement of gases through the stomata is restricted when the plant is under moisture stress.

The size of the root system is determined not only by the genetic character of the plant but also by the rooting conditions. The extent of root growth will be determined by soil physical conditions (see Chapter 5) and levels of soil moisture. Under high soil moisture conditions or around a drip emitter, root growth will be less than where there is not an excess of water present. A mature tomato plant will wilt if the plant is not able to draw sufficient water through its roots, a condition that can occur if the rooting medium is cool or the rooting zone is partially anaerobic (Carson, 1974). Also the size of the root system may be an important factor, but just how large the root system must be to ensure sufficient rooting surface for water absorption is a factor that is not known.

Under low moisture conditions surrounding the roots, there will be fewer flowers per truss, lower fruit will set if at 25% less than that needed, and blossom-end rot (BER) incidence will be high.

Soil moisture control in the field was obtained by Geraldson (1963, 1982) by maintaining the underlying water table over a raised plastic-covered bed. Today, the use of the plastic culture technique (Lamont, 1996) and drip irrigation (Clark and Smajstria, 1996a, 1996b; Hartz, 1996; Hartz and Hochmuth, 1996) provides the control needed to maintain the supply of water and essential plant nutrients at optimum levels. The drip irrigation technique (Nakayama and Bucks, 1986) is in wide use both in the field and in the greenhouse for supplying water to the plant at precise rates and times. In their book, Keller and Bliesner (1990) have reviewed the various types of irrigation systems including the drip method.

Geisenberg and Stewart (1986) gave the water requirement for field tomatoes as 2000–6600 m^3 ha^{-1} under normal air temperatures. In a greenhouse setting, a tomato plant in full fruit production will consume about 1 L of water per day. Rudich and Luchinisky (1986) indicated that solar radiation is the major determinant for water consumption, and at full canopy about 65% of the radiation reaching the crop is used to evaporate water. At 68°F (20°C), 585 cal m^{-2} are required to evaporate 1 cm^3 water. Based on this figure, one can calculate how much water would need to be applied by irrigation to replace what was lost by evaporation from the plant canopy.

Fruit yield and quality are factors that are affected by the amount of water available to the plant. Adams (1990) found in a peat bag system, restricting water to 80% or less based on the estimated requirement reduced yield by 4% due to smaller sized fruit but improved the flavor components of the fruit. Papadopoulos (1991) describes several organic mix bag systems for tomato production in which the major factor needing careful monitoring is moisture control.

3 Fruit Characteristics

"the ideal tomato, from the consumer's viewpoint, is one that is full size, vine ripened, unblemished, and characteristically at the red-ripe stage or anything near that stage"

CONTENTS

PHYSICAL

The tomato fruit is classified botanically as a berry, the size varying from small cherry types with only two divisions of the ovary (locules) (Figure 3.1) to large multilocular 'beefsteak' types (Figure 3.2). The number of locules defines the fruit type as follows:

Number of Locules	Fruit Type
Two	Cherry and plum or pear types (processing tomatoes)
Four to six	Commercial cultivars for fresh market
More than six	Large 'beefsteak' type for garden or greenhouse production (do not ship well, subject to cracking and irregularly shaped fruit)

QUALITY FACTORS

Consumers measure the quality of tomato fruit primarily by three factors, physical appearance (color, size, shape, defects, and decay), firmness, and flavor, factors that are discussed in this chapter. The nutritional characteristics of the tomato have gained interest because consumers are becoming more health conscious.

NUTRITIONAL CONTENT

There is growing public interest in bringing into the diet foods that can have a significant effect on bodily health; and that contain substantial levels of vitamins, minerals, and antioxidants. The tomato fruit has attracted considerable attention since the red pigment in the tomato fruit, lycopene, is an antioxidant; and the fruit also contains substantial quantities of vitamin A [red fruit containing on the average 1000 International Units (IU) per 100 g] and ascorbic acid (vitamin C ranging in content from 20–25 mg/100 g), and potassium (K) (200–210 mg/100 g). Most tomato varieties vary in soluble solids from 4.5–7.0%, with much of the soluble solids being fructose or glucose. Citric acid is the predominate acid in tomato juice and the pH of fruit is normally at or below 4.5.

The composition of tomato fruit as reported from different sources is given in Tables 3.1, 3.2, 3.3, and 3.4. An approximate comparison of the nutrient content by tomato fruit type is shown in Table 3.4. The vitamin contents of fruit are given in Tables 3.5 and 3.6.

Nitrogen fertilization can have a significant effect on the vitamin content of fruit as indicated by Mozafar (1993):

Vitamin	Nitrogen Effect
Ascorbic acid	Decrease/increase (mixed effect)
Carotene	Increase

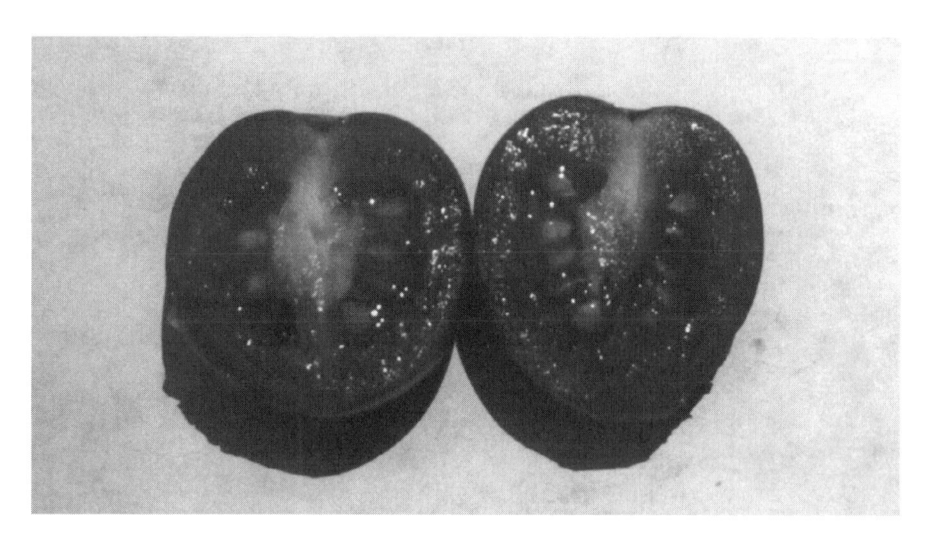

FIGURE 3.1 Two-locule tomato fruit.

FIGURE 3.2 Cross section of two 'beefsteak'-type tomatoes, round in shape (top) and oval (bottom).

Based on average fruit composition, the daily consumption of a medium-sized tomato weighing 8 oz would supply the following percentage of the recommended daily adult requirement (RDAR) as set by the Food and Drug Administration (FDA) in the Nutrition Labeling and Education Act (NLEA) of 1990:

Nutrient	DAR (%)
Vitamin C	47
Vitamin A	22
Thiamin	10
Riboflavin	6
Niacin	5
Fiber	10
Protein	4
Sodium (Na)	100
Potassium (K)	17
Iron (Fe)	6

The changes that occur in the tomato fruit with ripening have been listed by Grierson and Kader (1986) as:

- Degradation of starch and production of glucose and fructose
- Loss of chlorophyll
- Synthesis of pigments such as β-carotene and lycopene
- Increase in soluble pectins resulting from wall softening and degradation
- Production of flavor and aroma compounds
- Increase in ratio of citric acid and malic acid
- Increase in glutamic acid
- Breakdown of the toxic alkaloid α-tomatine

Lycopene

Lycopene is being called the "world's most powerful antioxidant," which can help to prevent the development of many forms of cancer, the effect varying with sex and type of cancer. The best known source of lycopene is in cooked tomatoes or tomato products since lycopene is released from the tomato on cooking. Raw tomatoes have about one fifth the lycopene content as that found in cooked tomato products. However, tomato, whether raw or cooked, is still the best source for this antioxidant, while watermelon and grapefruit are also good sources. Scientists have bred new tomato varieties that have a high lycopene content in the fruit so that lycopene can be extracted to make natural supplements; one currently available is called Lyc-O-Mate. Lycopene as well as carotenes are not synthesized in the fruit when the air temperature is less than 50°F (10°C) or above 85°F (29.4°C).

Salicylates

Another constituent found in the tomato fruit is salicylate, an aspirin-like substance that has been found to reduce the risk of heart disease.

pH

The range of pH for the tomato fruit is between 4.0 and 4.5; the lower the pH, the greater is the so-called "tartness," a factor by which some consumers judge the

TABLE 3.1
Composition of Green and Ripe Tomato Fruit

Constituents	Green	Ripe
	%	
Water	93	94
Fat	0.2	0.2
	(amount per 100 g)	
Protein, g	1.2	0.9
Carbohydrates, g	5.1	4.3
Fiber, g	0.5	0.8
Iron, g	0.5	0.5
Calcium, mg	13	7
Phosphorus, mg	28	23
Sodium, mg	13	8
Potassium, mg	204	207
Thiamin, mg	0.06	0.06
Riboflavin, mg	0.04	0.05
Niacin, mg	0.50	0.60
Ascorbic acid, mg	23.4	17.6
Vitamin B_6, mg	—	0.05
Energy, kcal	24	19
Vitamin A (IU)	1133	7600

Note: IU, international units.

Source: Lorenz, O.A., and D.N. Maynard. 1988. *Knott's Handbook for Vegetable Growers*. 3rd ed., John Wiley & Sons, New York.

quality of the tomato fruit. The average pH range for most fruit lies between 4.3 and 4.4. In *All About Tomatoes* (Ray, 1976), a table titled "tomato pH—past and present" suggests that newer varieties produce fruit of lesser pH:

Year of Introduction	Number of Varieties of Breeding Lines	Mean pH
Before 1950	49	4.29
1950–1959	26	4.34
1960–1969	73	4.35
1970–1976	96	4.34

It is doubtful that there has been a substantial change in the pH of fruit since 1976 because current pH measurements of fruit range between 4.30 and 4.40.

TABLE 3.2
Tomato Fruit Composition from the U.S. Department of Agriculture

Nutrient	Content (135-g Sample)
Protein, g	1
Fat, g	Trace
Carbohydrate, g	6
Calcium, mg	16
Phosphorus, mg	33
Iron, mg	0.6
Potassium, mg	300
Thiamin, mg	0.07
Riboflavin, mg	0.05
Niacin, mg	0.9
Ascorbic acid, mg	28[a]
Food energy, cal	25
Vitamin A (IU)	1110
Water, %	94

Note: IU, international units.

[a] Average year-round value: marketed November–May, 12 mg; June–October, 32 mg.

Source: Adams, C.F. and M. Richardson. 1977. *Nutritive Value of Foods*. USDA-ARS Home and Garden Bulletin Number 72, U.S. Government Printing Office, Washington, D.C.

The relationship between the pH and solids content (mainly sugars) of the tomato fruit is also a significant factor in its perceived flavor. The degree of ripeness is also a factor that affects the pH of the fruit. The role that factors other than genetics have on fruit flavor external to the plant, such as climate, soil, and cultural practices, may be equally significant factors, factors that are just beginning to be more carefully investigated.

When canning, the pH of the canned product will determine the safety of the final product. To ensure adequate acidity, 1/4 teaspoon of citric acid or 1 tablespoon of lemon juice should be added per pint of tomato product prior to canning.

SUN-DRIED FRUIT

When the tomato fruit is sun dried, there is a loss of some nutrients, particularly vitamin C (ascorbic acid). A half ounce of dried fruit contains 7 mg of vitamin C while a medium-sized fresh tomato fruit contains 24 mg, which is 40% of the RDAR. This same fresh fruit contains 24 cal and 11 mg of sodium (Na), while 1 oz of dried tomato fruit contains 73 cal and 5194 mg of Na, increases that could significantly impact the diet of those wanting to restrict their calorie and Na intake.

TABLE 3.3
Nutritive Value of Fresh Tomatoes

Constituent	Composition (100-g Edible Portion)
Protein, g	1.1
Fat, g	0.2
Carbohydrate, g	4.7
Fiber, g	0.5
Ash, g	0.5
Sodium, g	3
Calcium, mg	13
Phosphorus, mg	27
Iron, mg	0.5
Potassium, mg	244
Thiamin, mg	0.06
Riboflavin, mg	0.04
Niacin, mg	0.7
Ascorbic acid, mg	23[a]
Food energy, cal	22
Water, %	93.5
Vitamin A (IU)	900

Note: IU, international units

[a] Average year-round value: marketed November–May, 10 mg/100 g; June–October, 26 mg/100 g.

Source: Anon. 1978. *Nutritive Values of Fruit and Vegetables.* United Fresh Fruit and Vegetable Association, Alexandria, VA.

GOVERNMENT QUALITY FACTORS

Quality standards for grades for processing vegetables in the United States have been established as published by Kader (1992); the quality standard characteristics are

Fruit Type	Year	Characteristics
Tomato	1983	Firmness; ripeness (color as determined by a photoelectric instrument); and freedom from insect damage, freezing, mechanical damage, decay, growth cracks, sunscald, gray wall, and blossom-end rot (BER)
Green	1950	Firmness, color (green), and freedom from decay and defects (growth cracks, scars, catfacing, sunscald, disease, insects, or mechanical damage)
Italian type for canning	1957	Firmness, color uniformity, and freedom from decay and defects (growth cracks, sunscald, freezing, disease, insects, or mechanical injury)

TABLE 3.4
Approximate Nutrient Composition between Fruit Type

Constituent	Tomato	Cherry Tomato
	%	
Water	93.5	93.2
Calories	22	22
CHO	4.75	4.9
Protein	1.05	1.00
Fat	0.20	0.20
Fiber	0.55	0.40
Ash	0.50	0.70
Calcium (Ca)	12	29
Phosphorus (P)	26	62
Potassium (K)	244	—
Sodium (Na)	3	—
Magnesium (Mg)	14	—
Iron (Fe)	0.5	1.7
Vitamin A	900	2000
Vitamin C	25	50
Vitamin B_1	0.06	0.05
Vitamin B_2	0.04	0.04
Niacin	0.7	—

Note: CHO, carbohydrate.

TABLE 3.5
Vitamin Content of Edible Green and Ripe Tomato Fruit

Vitamin	Amount per 100 g Edible Portion	
	Green	Ripe
Vitamin A (IU)	642	1133
	mg	
Thiamin	0.06	0.06
Riboflavin	0.04	0.05
Niacin	0.50	0.60
Ascorbic acid	23.40	17.60
Vitamin B_6	0.05	0.38

Note: IU, international units.

TABLE 3.6
Vitamin Content of Ripe Tomato Fruit

Vitamin	Range
A (β-carotene)	900–1271 IU[a]
B_1 (thiamin)	50–60 μg
B_2 (riboflavin)	20–50 μg
B_3 (pantothenic acid)	50–750 μg
B_6 complex	80–110 μg
Nicotinic acid (niacin)	500–700 μg
Folic acid	6.4–20 μg
Biotin	1.2–4.0 μg
Vitamin C	15,000–23,000 μg
Vitamin E (α-tocopherol)	40–1,200 μg

Note: Range of values per 100 g of fruit.

[a] IU (international units) = 0.6 μg β-carotene.

Source: Davies, J.N. and G.E. Hobson. 1981. *CRC Crit. Rev. Food Sci. Nutri.* 15:205–280.

These quality standards are used for processing tomato fruit and do not apply to fresh market fruit.

Fresh tomato fruit standards based on the United States standards for grades of fresh tomatoes (effective December 1, 1973; amended November 29, 1973, February 1, 1975, and April 15, 1976) (Anon., 1976) follow.

Size

The size of tomatoes packed into any type container is specified according to the size designation set forth as:

	Diameter (in.)		Diameter (mm)	
Size Designations	Minimum[a]	Maximum[b]	Minimum[a]	Maximum[b]
Extra small	$1\frac{28}{32}$	$2\frac{4}{32}$	48	54
Small	$2\frac{4}{32}$	$2\frac{9}{32}$	54	58
Medium	$2\frac{9}{32}$	$2\frac{17}{32}$	58	64
Large	$2\frac{17}{32}$	$2\frac{28}{32}$	64	73
Extra large	$2\frac{28}{32}$	$3\frac{14}{32}$	73	88
Maximum large	$3\frac{15}{32}$	—	88	—

[a] Will not pass through a round opening of the designated diameter when the tomato is placed with the greatest transverse diameter across the opening.

[b] Will pass through a round opening of the designated diameter in any position.

Color Classification

The following terms may be used when specified in connection with the grade statement, in describing the color as an indication of the stage of ripeness of any lot of mature tomatoes of a red-fleshed variety:

Green—The surface of the tomato is completely green in color. The shade of green color may vary from light to dark.

Breakers—There is definite break in color from green to tannish-yellow, pink, or red on not more than 10% of the surface.

Turning—More than 10% but not more than 30% of the surface in the aggregate shows a definite change in color from green to tannish-yellow, pink, red, or a combination of these.

Pink—More than 30% but not more than 60% of the surface in the aggregate shows pink or red color.

Light red—More than 60% but not more than 90% of the surface in the aggregate shows pinkish-red or red color, provided that not more than 90% of the surface is red.

Red—More than 90% of the surface in the aggregate shows red color.

Tolerance

To allow for variations incident to proper grading and handling in each of the grades, the following tolerances by count are provided as specified:

U.S. No. 1 Grade

Basic requirements include

1. Similar varietal characteristics
2. Mature
3. Not overripe or soft
4. Clean
5. Well developed
6. Fairly well formed
7. Fairly smooth

The fruit should be free from:

1. Decay
2. Freezing injury
3. Sunscald
4. Damage by any other cause

U.S. No. 2 Grade

Basic requirements include

1. Similar varietal characteristics
2. Mature

3. Not overripe or soft
4. Clean
5. Well developed
6. Reasonably well formed
7. Not more than slightly rough

The fruit should be free from:

1. Decay
2. Freezing injury
3. Sunscald
4. Serious damage by any other cause

U.S. No. 3 Grade

Basic requirements include

1. Similar varietal characteristics
2. Mature
3. Not overripe or soft
4. Clean
5. Well developed
6. May be misshapen

The fruit should be free from:

1. Decay
2. Freezing injury
3. Serious damage by sunscald or any other cause

In the grade designation, defects are specified at the point of shipment, and defects en route or at destination.

Classification of Defects

Defect factors are classified as to level, *damage*, *serious damage*, or *very serious damage;* and the factors include cuts and broken skins, puffiness, catfacing, scars, growth cracks, hail injury, and insect injury.

Definitions

Various definitions are given that define the factors used to classify fresh tomato fruit; a brief description of each of these follows:

Similar varietal characteristics—Tomatoes are alike as to firmness of flesh and shade of color. For example, soft-fleshed, early maturing varieties are not mixed with firm-fleshed, midseason or late varieties, or bright red varieties mixed with varieties having a purplish tinge.

Mature—Tomato has reached the stage of development that will ensure a proper completion of the ripening process, and that the contents of two or more seed cavities have developed a jellylike consistency and the seeds are well developed.

Soft—Tomato yields readily to slight pressure.

Clean—Tomato is practically free from dirt or other foreign material.

Well developed—Tomato shows normal growth. Tomatoes that are ridged and peaked at the stem end, contain dry tissue, and usually contain open spaces below the level of the stem scar are not considered well developed.

Fairly well formed—Tomato is not more than moderately kidney shaped, lopsided, elongated, angular, or otherwise moderately deformed.

Fairly smooth—Tomato is not conspicuously ridged or rough.

Damage—Damage means any specific defect (cuts and broken skins, puffiness, catfacing, scars, growth cracks, hail, and insect injury) or any one of these defects, any other defect, or any combination of defects, which materially detracts from the appearance, or the edible or marketing quality of the tomato.

Reasonably well formed—Tomato is not decidedly kidney shaped, lopsided, elongated, angular, or otherwise decidedly deformed.

Slightly rough—Tomato is not decidedly ridged or grooved.

Serious damage—This means specific defects (cuts and broken skins, puffiness, catfacing, scars, growth cracks, hail, and insect injury) or an equally objectionable variation of any one of these defects, any other defect, or any combination of defects, which seriously detracts from the appearance, or the edible or marketing quality of the tomato.

Misshapen—Tomato is decidedly kidney shaped, lopsided, elongated, angular, or otherwise deformed, provided that the shape is not affected to an extent that the appearance or the edible quality of the tomato is very seriously affected.

Very serious damage—This means any specific defect (cuts and broken skins, puffiness, catfacing, scars, growth cracks, hail, and insect injury) or an equally objectionable variation of any one of these defects, any other defect, or any combination of defects, which seriously detracts from the appearance, or the edible or marketing quality of the tomato.

RIPENING AND COLOR DEVELOPMENT

Fruit ripening is a complex biochemical process in which the chlorophyll and starch content of the fruit decreases and the softening enzyme polygalacturonase and lycopene increases (Grierson and Kader, 1986). With the beginning of the ripening process, respiration [synthesis and release of carbon dioxide (CO_2)] and generation of ethylene (C_2H_4) increases, peaking after about 10 days and then declining.

The stages of tomato ripening and color development are used to identify tomato fruit, stages, and their corresponding fruit characteristics, which are given in Table 3.7.

The "breaker" stage of fruit development is the most commonly chosen stage for picking fresh market fruit that is to be shipped some distance; while locally marketed fruit may stay on the vine until the light red to red stage is reached. Fruit

TABLE 3.7
Stages of Tomato Fruit Ripening and Color Development for Red-Fruited Cultivars

Harvest Stages	Days from Mature Green at 68°F (20°C)	Fruit Characteristics
Immature green		Fruit still enlarging, dull green, lacks skin luster, gel not well formed; seed easily cut through when fruit is sliced; immature seed not germinating, and fruit not coloring properly
Mature green	0	Bright to whitish green; well rounded, skin with waxy gloss; seeds embedded in gel and not easily cut when fruit is sliced; seeds mature and can germinate; fruit ripening under proper conditions
Breaker	2	Showing pink color at blossom end; internally the placenta pinkish
Turning	4	Pink color extending from blossom end, covering 10–30% of fruit
Pink	6	Pink to red color covering 30–60% of fruit
Light red	8	Pink to red color covering 60–90% of fruit
Red	10	Red color at least 90% of fruit

Source: Rubatzky, V.E. and M. Yamaguchi. 1997. pp. 533–552. In: V.E. Rubatzky and M. Yamaguchi (eds.), *World Vegetables: Principles, Production, and Nutritive Values.* Chapman & Hall, New York.

that is picked in the mature green stage will have a very long keeping period, but there will be a significant loss in color and flavor in the finally marketed fruit.

Ripening of mature green and breaker stage fruit can be hastened by treatment with ethylene (C_2H_4) at 100–150 mg L^{-1} (ppm) in the storage atmosphere, mature green developing a red color 5 to 7 days at 65–68°F (18.3–20°C), which can be either increased or decreased by higher or lower temperatures, respectively (Peet, 1996b).

If green fruit is refrigerated at 42°F (5.5°C) (typical refrigerator temperature), the ripening enzymes in the fruit are inactivated, and it will never ripen.

DAYS TO MATURITY

The approximate time from pollination to market maturity under warm growing conditions varies from 35–60 days (Maynard and Hochmuth, 1997), with the days to maturity depending on the stage of maturity when harvested:

Market Stage	*Days to Market*
Mature green	35–45
Red ripe	45–60

STORAGE AND RIPENING

The relative perishability and potential storage life of fresh tomato fruit in air at near optimum storage temperature and relative humidity is about 2 weeks for ripe fruit and 2–4 weeks for partially ripe fruit. For the fresh fruit market, the time expected between harvest and market determines the degree of fruit ripening at harvest. It has been observed that the keeping quality of greenhouse-grown fruit is longer than that for field grown, probably because the greenhouse fruit is free of soilborne organisms.

The optimum storage temperature for ripe fruit is between 45 and 50°F (7.2 and 10°C) and the relative humidity is from 85–96%. Mature green fruit can be stored at 55–60°F (12.7–15.5°C) for several days without significant quality losses. To maintain the quality of red fruit, low temperature exposure must be avoided. Storage time based on fruit type, temperature, and relative humidity is shown in Table 3.8.

TABLE 3.8
Storage Time Based on Fruit Type, Temperature, and Relative Humidity

Fruit Type	Temperature [°F (°C)]	Relative Humidity (%)	Storage Time (weeks)
Firm ripe	46–50 (7.7–10)	90–95	1–3
Mature green	55–70 (12.7–21)	90–95	4–7

TABLE 3.9
Respiration Rates of Mature and Ripening Fruit at Various Temperatures

Fruit Type	Temperature [°F (°C)]	Respiration Rate (mg kg^{-1} hr^{-1} of CO_2)
Mature green	50 (10)	12–18
	59–60 (15–15.5)	16–28
	68–70 (20–21)	28–41
	77–80 (25–26.6)	35–51
Ripening	50 (10)	13–16
	59–60 (15–15.5)	24–29
	68–70 (20–21)	24–44
	77–80 (25–26.6)	30–52

With varying storage temperature and fruit type, the respiration rate varies, as shown in Table 3.9.

TABLE 3.10
Controlled Conditions for the Long-Term Storage of Mature Green and Partially Ripe Fruit

Fruit Type	Temperature [°F (°C)]	Oxygen (%)	Carbon Dioxide (%)	Benefit
Mature-green	54–68 (12.2–20)	3–5	0–3	Potential benefit good, limited commercial use
Partially ripe	46–54 (7.7–12.2)	3–5	0–5	Potential benefit good, limited commercial use

TABLE 3.11
Susceptibility to Chilling Temperature

Fruit	Lowest Safe Temperature [°F (°C)]	Appearance When Stored 32°F to Safe Temperature
Ripe	45–50 (7.2–10)	Water soaking and softening decay
Mature green	55 (12.7)	Poor color when ripe, alternaria rot

Source: Lorenz, O.A. and D.N. Maynard. 1988. *Knott's Handbook for Vegetable Growers*. 3rd ed., John Wiley & Sons, New York.

The use of controlled atmosphere for tomato fruit storage is not generally practiced, although it would be of considerable benefit for the long-term storage of fruit. The control conditions are shown in Table 3.10.

The lowest safe temperature by fruit type, temperature, and appearance when stored between 32°F (0°C) and the safe temperature are shown in Table 3.11.

Tomato fruit should never be put on ice. If a tomato fruit is frozen, the fruit will become water soaked and will soften on thawing; and if partially frozen, the margins between healthy and dead tissue are distinct, especially in green fruits. The effects of chilling injury temperatures on tomato fruit characteristics are

Type of fruit	Fruit characteristics
Ripe fruit	Water soaking and softening, decay
Mature green	Poor color when ripe, alternaria rot
Freezing injury	Small water-soaked spots or pitting on the surface; injured tissues appearing tan or gray and giving off an objectionable odor

Fresh cooked, or canned and opened tomato fruit can be stored in the refrigerator at 45°F (7.2°C) for 4–5 days; cooked tomato dishes, in the refrigerator freezer compartment for 2–3 months; and cooked dishes, in a freezer at 32°F (0°C) for 1 year. Unopened canned tomato can be stored on the kitchen shelf for 1 year.

HOME LONG-TERM STORAGE

For those wanting to extend the season into the early winter months, tomato fruit collected in the very early breaker stage (see Table 3.7) in the fall can be stored for 3–4 months for eating during the holiday period in November and December. Not all varieties store equally well; yellow-fruited varieties, in general, store better than red-fruited varieties.

In an article in the *Organic Gardening* magazine (Anon., 1997b), the longer keeping varieties (in order of taste) were

Yellow varieties	Dwarf Gold Treasure, Mountain Gold, and Winter Gold
Red varieties	Flavor More, Sheriff, Winter Red

Cebenko (1997) has outlined the requirements and procedure for storing home garden tomatoes picked at most stages of development. His procedure is as follows:

- Any fruit that has been bruised should not be stored because such fruit will rot before ripening.
- Place fruit in a newspaper-lined box with newspaper placed between each individual fruit but one-layer deep, with the fruit uncovered on the top (do not wrap the fruit in newspaper).
- Put the box in a dark, cool [35–45°F (1.6–7.2°C)], and humid (70% or more) place.
- Inspect the fruit once a week and remove ripe fruit or any fruit beginning to decay (check the stem and blossom end where decay is likely to begin).
- Speed of ripening can be controlled by temperature; the higher the temperature, the faster the ripening takes place.

FACTORS AFFECTING FRUIT QUALITY

Fruit quality is significantly affected by stage of ripeness (see Table 3.7) when removed from the plant, number of times handled, and storage temperature and time (Grierson and Kader, 1986; Sargent et al., 1997). The longer the fruit remains on the plant, the more flavorful the fruit is. Less handling reduces the incidence of bruising, and some have suggested that flavor is reduced with increased handling. It is frequently observed that tomato fruit not ripened on the plant does not have the same flavor and aroma as fruit that has developed its red color on the plant.

Harvested fruit that is bruised in picking and transport will release ethylene (C_2H_4) that will hasten the ripening process. In studies by Sargent et al. (1997), they found that harvest maturity, storage temperatures [68°F (20°C)], and internal bruising negatively affected tomato flavor and quality. Moretti et al. (1997) found that "impact bruising" reduced fruit quality, affecting both its chemical and physical characteristics, similar to findings reported by Grierson and Kader (1986).

COMMONLY OCCURRING FRUIT DISORDERS

There are ten commonly occurring fruit disorders that either are genetic in origin or have their origin in production or handling procedures.

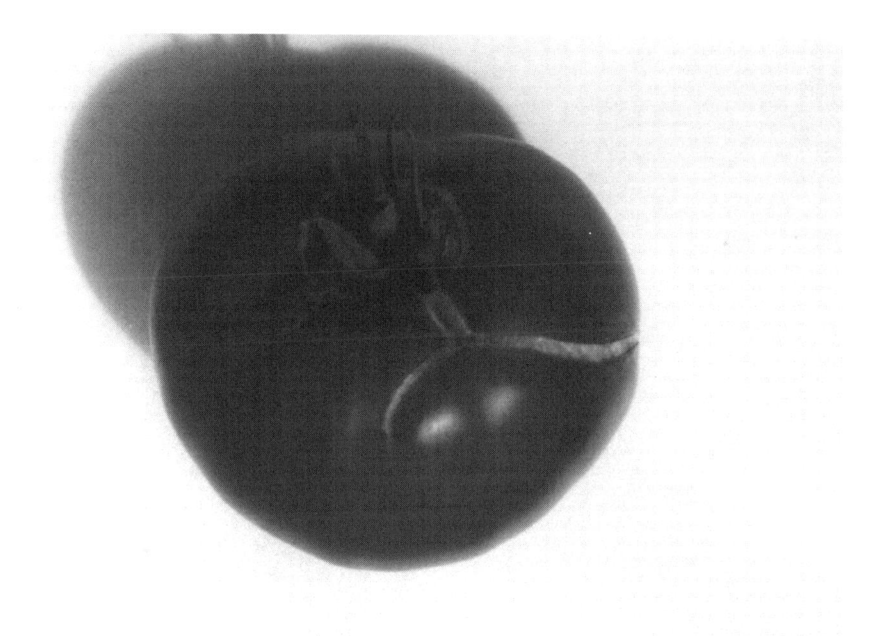

FIGURE 3.3 Cracked tomato fruit due to sudden change in water content that breaks the skin.

Cracking

As the fruit begins to ripen, the skin may crack, particularly during warm wet periods if there has been a preceding dry spell. Fruit cracking occurs when there is a rapid net influx of water and solutes into the fruit at the same time that ripening or other factors reduce the strength and elasticity of the tomato skin (Peet, 1992). There are varietal differences because some varieties have excellent crack resistance. Cracking can be minimized by selecting a variety that has resistance to cracking as well as maintaining a consistent soil moisture to avoid periods of plant moisture stress (Abbott et al., 1986; Peet, 1992; Peet and Willits, 1995). Koske et al. (1980) found that ground bed heating did not affect the incidence of cracking.

Cracking may be concentric encircling the stem end of the fruit or radial starting at the stem end and extending sometimes to the blossom end (Figure 3.3). The percentage of cracking increases with number of fruit per cluster, and more frequent watering also increases the incidence of cracking as was found by Peet and Willits (1995).

Anatomic characteristics most frequently associated with fruit cracking given by Peet (1992) are

- Large fruit size
- Low skin tensile strength or low skin extensibility at the turning to the pink stage of ripeness
- Thin skin
- Thin pericarp
- Shallow cutin penetration

FIGURE 3.4 Catfaced tomato fruit due to incomplete pollination of all the seed cells.

- Few fruits per plant
- Fruit not shaded by foliage

Catfacing

A catfaced fruit is misshapen due to abnormal development that begins at the time of flowering, believed to be due, in part, to cool temperatures [less than 55°F (12.8°C)] and cloudy weather at the time of flowering and fruit set. Although catfacing is usually a specific disorder in terms of fruit appearance, any misshapen fruit due to incomplete pollination may be also identified as catfacing (Figure 3.4).

Misshapen Fruit

Common abnormalities are pointed fruits with an elongated blossom end, puffy fruit in which air spaces have developed, or lack of round smoothness that may be similar to catfacing. The cause is usually due to low air temperature [less than 55°F (12.7°C)] and cloudy weather that interferes with the growth of the pollen tubes and normal fertilization of the ovary.

Puffiness

Puffiness most frequently occurs in early-harvested fruit caused by a variety of conditions, such as high [>90°F (32.2°C)] and low [<58°F (14.4°C)] air temperatures, low light, excessive nitrogen (N) fertilization, or heavy rainfall, in which one or more seed cavities are empty (Figure 3.5).

FIGURE 3.5 Tomato fruit that would be classed as puffiness. (Note the lack of seeds in the right locule.)

Blossom-End Rot

The blossom end of the fruit first turns light brown and then black as the cells at the blossom-end decay (Figure 3.6). This disorder, frequently referred to by its acronym BER, has several causes, the most common being calcium (Ca) deficiency coupled with moisture stress, which is probably the triggering mechanism (Pill et al., 1978). Grierson and Kadar (1986) found that when the dry weight Ca content of the fruit was 0.12%, BER did not occur, but when 0.08%, BER did occur. However, a number of other stresses when combined with other types of physiological stress will result in BER-affected fruit. The incidence of BER increases significantly with fruit thinning (DeKock et al., 1982). Although fruit thinning will increase fruit size, BER occurrence will then increase significantly in newly developing fruit; this increase in occurrence is due to the excessive supply of hormones from the roots to the developing fruit. High availability of ammonium (NH_4) as the N source will significantly increase the occurrence of BER (Wilcox et al., 1973; Hartman, et al., 1986).

Sunscald (Solar Injury)

Sunscald occurs on green (most sensitive) and ripening (less sensitive) fruit from exposure to direct sunlight for long periods of time. On solar-injured or sunburned (solar yellowing) fruit, affected areas on the fruit become whitish, translucent, and thin walled; and a netted appearance may develop. Mild solar injury might not be noticeable at harvest, but becomes more apparent after harvest as uneven ripening. Direct sun exposure will result in a significant increase in fruit temperature that

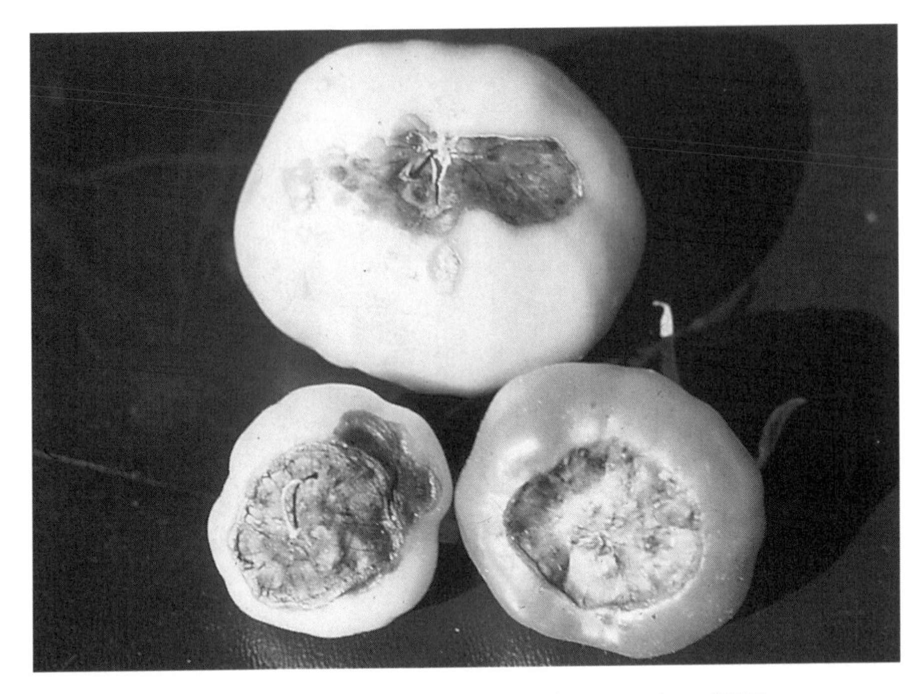

FIGURE 3.6 Tomato fruits affected by blossom-end rot (BER).

affects its development and quality (Grierson and Kadar, 1986). Foliage cover is the most effective way to reduce sunscalding of fruit.

Green Shoulders

From the exposure to direct sunlight, the shoulder of the fruit will remain green. Foliage cover, which is frequently a genetic factor, is one means of reducing the occurrence of this problem.

Russeting

Russeting is characterized by a brownish scarring of the surface of the fruit, giving it a rough, clouded looking appearance. This is believed to be due to very humid greenhouse conditions.

Anther Scarring

Anther scarring is the presence of a long scar along the blossom end of the fruit that is probably due to some early injury to the flower, although the exact cause is not known (Figure 3.7).

Blotchy Ripening

Uneven ripening of the fruit with various areas of green to yellowish green on the fruit is called blotchy ripening. The cause is not exactly known but is frequently associated with low potassium (K) and high nitrogen (N) nutrition of the plant.

FIGURE 3.7 Tomato fruit showing anther scarring.

FLAVOR

The tomato fruit is mostly water with about 5–7% of the fruit being solids, with most fruit solids content being closer to 5% than to 7%. Of the solids content, about half is composed of sugars and one eighth, acids. Davies and Hobson (1981) reported on the composition of ripe tomato fruit as:

Constituent	Dry Matter (%)
Sugars	
Glucoses	22
Fructoses	25
Sucroses	1
Alcohol insoluble solids	
Proteins	8
Pectic substances	7
Hemicelluloses	4
Celluloses	6
Organic acids	
Citric acids	9
Malic acids	4
Minerals	
(mainly K, Ca, Mg, P)	8
Others	
Lipids	2
Dicarboxylic amino acids	2
Pigments	0.4
Ascorbic acids	0.5
Volatiles	0.1
Other amino acids, vitamins, and polyphenols	1.0

Fruit flavor is a major consumer demand and one that attracts much attention. Peet (1996d) has described fruit flavor based on acidity (low pH) and sugar content of the fruit as follows:

Acidity	Sugar Content	Flavor
High	High	Good
High	Low	Tart
Low	High	Bland
Low	Low	Tasteless

Stevens et al. (1977) found that fructose and citric acid were more important to sweetness and sourness rather than glucose and malic acid, and pH was a better objective measure of sourness than titratable acidity. They also were able to identify 11 volatile compounds with only 3 of these compounds being more important to the "tomato-like" character.

In general, the longer the fruit remains on the plant, the better the flavor as sensed by the consumer, although this not entirely true. Fruit that is harvested at the breaker stage and then properly handled can be flavorful (Grierson and Kader, 1986; Sargent et al., 1997). The term vine ripe may be put on fruit packaging to attract consumer attention because this term, in the minds of many consumers, is considered to be related to flavor.

Stevens (1986) has suggested that the composition of tomato fruit can be markedly improved through breeding and genetic engineering techniques, listing those composition factors that are currently in fruit with what is possible based on known genetic resources.

	Fresh Market		Processing	
Ingredient	**Present**	**Potential**	**Present**	**Potential**
	(%)	(%)	(%)	(%)
Total solids	5.8	7.5	5.7	7.5
Reducing sugars	3.2	4.3	2.7	3.9
Alcohol insoluble solids	0.7	1.2	1.2	1.6
Total acids	0.8	0.9	0.7	0.9
	$\mu g\ g^{-1}$	$\mu g\ g^{-1}$	$\mu g\ g^{-1}$	$\mu g\ g^{-1}$
Carotenoids				
Lycopene	40	80	48	80
β-Carotene	5	10	5	10

FRUIT COLOR

Normally one associates red as the normal tomato fruit color, although fruit at maturity can be either pink, various shades of yellow, or even green. For many years, the greenhouse tomatoes grown in the Cleveland, OH area were pink so that they could be distinguished from other sources of tomatoes in the marketplace. Today, there seems to be increased interest in yellow tomatoes (Mattern, 1996) as well as fruit of other shades of yellow-red and yellow-green (Weaver, 1998). However, red is still the major fruit color for most of the commonly grown varieties.

FREEDOM FROM PEST CHEMICALS

Various studies have been made to determine at what level various chemicals, whether herbicides, insecticides, or fungicides, used in the production of tomato fruit still remain either on the surface or within the tomato itself when brought to the marketplace. Most studies have focused on a variety of fruits and vegetables and not tomato alone. A study conducted by the National Cancer Institute of Canada, which examined fruits and vegetables for pesticide and fungicide residue content, found that 64–85% of the crops sampled had no detectable residues and more than 98% were below legal limits for residues. The results of this study, not dissimilar from other studies conducted by various United States agencies, indicated that most vegetables are free of pest chemicals, or that if levels do exist, they are well below the legal limits set for human consumption.

However, for some consumers, any level of residue found on the tomato fruit would be unacceptable; therefore, the designation on a tomato fruit as being "pesticide free," meaning that no pesticides were used in the production of the fruit, could be a significant marketing factor.

WASHING FRESH FRUIT FROM THE FIELD

After harvest, the fruit is water-washed to remove soil and dust particles. To minimize disease spread, the washing conditions are

- Keeping the fruit at one layer in the washing water
- Removing fruit from the water after 2 min
- Chlorinating the water, maintaining a free chlorine level of 300 ppm
- Warming the wash water to 10°F (–12.2°C) above the fruit temperature

FRUIT PACKAGING

Tomato fruit is brought into the marketplace in a number of different forms as has been discussed by Schales (1985):

- As loose fruit in a box (Figure 3.8)
- In a plastic tube, number of fruit determined by size of the fruit (Figure 3.9)
- In plastic clamshells of one or more compartments (Figure 3.10)
- Cluster tomatoes with the fruit left on the truss and placed in a plastic bag (Figure 3.11)

The calyx is normally removed if the fruit is not packaged to keep from damaging other fruit, or the calyx is kept on the fruit to make the fruit look as if it has just been taken from the plant (see Figure 3.12).

FIGURE 3.8 Loose tomato fruit placed in a suitably designed box.

FIGURE 3.9 Tomatoes packaged in a plastic tube.

FIGURE 3.10 Cherry tomato fruit placed in a plastic clamshell container.

FIGURE 3.11 Cluster tomatoes placed in a plastic bag.

FIGURE 3.12 Tomato fruit without (left) and with (right) the calyx.

USEFUL UNITS

Consumers are always looking for weight and volume information to help them determine how many pieces of fruit make a pound, etc. The following weight measurement information is given to provide this assistance:

- One pound of tomatoes consists of one very large, two large, three medium (2 × 2½ inches), or four small tomatoes.
- An 8-oz can equals 1 cup of cooked tomatoes.
- A 12-oz can equals 1½ cups of cooked tomatoes.
- A 16-oz can equals 2 cups of cooked tomatoes.
- A 28-oz can equals 3½ cups of cooked tomatoes.
- A 46-ounce can of juice equals 5¾ cups.
- One bushel of tomatoes (50 lb) will make 20 qt of cooked tomatoes.

4 Plant Nutrition

CONTENTS

ESSENTIAL ELEMENTS

There are 16 elements that have been identified as being essential for the normal growth and development of all plants, the requirements of essentiality having been established by Arnon and Stout (1939). The major texts on plant nutrition are the books by Epstein (1972), Marschner (1986), Mengel and Kirkby (1987), and Glass (1989), and a basic manual on plant nutrition by Jones (1997a) as well as a video on the subject by Jones (1993a).

For 13 of the 16 essential elements, their approximate average concentration in the plant is shown in Table 4.1.

TABLE 4.1
Average Concentration of Mineral Nutrients in Plant Dry Matter that Is Sufficient for Adequate Growth

Element	Abbreviation	µmol g^{-1} dry wt	mg kg^{-1} (ppm)	Percentage (%)	Relative Number of Atoms
Molybdenum	Mo	0.001	0.1	—	1
Copper	Cu	0.10	6	—	100
Zinc	Zn	0.30	20	—	300
Manganese	Mn	1.0	50	—	1,000
Iron	Fe	2.0	100	—	2,000
Boron	B	2.0	20	—	2,000
Chlorine	Cl	3.0	100	—	3,000
Sulfur	S	30	—	0.1	30,000
Phosphorus	P	60	—	0.2	60,000
Magnesium	Mg	80	—	0.2	80,000
Calcium	Ca	125	—	0.5	125,000
Potassium	K	250	—	1.0	250,000
Nitrogen	N	1,000	—	1.5	1,000,000

Source: Epstein, E. 1965. pp. 438–466. In: J. Bonner and J.E. Varner (Eds.), *Plant Biochemistry*. Academic Press, Orlando, FL.

The form for uptake and general functions for the 16 essential elements in plants are given in Table 4.2.

TABLE 4.2
The Essential Elements, Their Form for Uptake, and Functions in the Plant

Essential Element	Form for Uptake	Functions in the Plant
C, H, O, N, S	Ions in solution (HCO_3^-, NO_3^-, NH_4^+, SO_4^{2-}), or gases in the atmosphere (O_2, N_2, SO_2)	Major constituents of organic substances
P, B	Ions in solution (PO_4^{3-}, BO_3^{3-})	Energy transfer reactions and carbohydrate movement
K, Mg, Ca, Cl	Ions in solution (K^+, Mg^{2+}, Ca^{2+}, Cl^-)	Nonspecific functions, or specific components of organic compounds, or maintaining ionic balance
Cu, Fe, Mn, Mo, Zn	Ions or chelates in solution (Cu^{2+}, Fe^{2+}, Mn^{2+}, MoO^-, Zn^{2+})	Enable electron transport and catalysts for enzymes

Source: Mengel, K. and E.A. Kirkby. 1987. *Principles of Plant Nutrition*, 4th ed. International Potash Institute, Bern, Switzerland.

STRUCTURAL ELEMENTS

When chlorophyll-containing plant tissue is in the presence of light, three of the essential elements, carbon (C), hydrogen (H), and oxygen (O), are combined in the process called *photosynthesis* to form a carbohydrate. Carbon dioxide (CO_2) is from the air, and water (H_2O) is taken up through the roots. In the photosynthetic process, a water molecule is split and combined with CO_2 to form a carbohydrate while a molecule of oxygen (O_2) is released. Since tomato is a C3 plant, the first product of photosynthesis is a 3-carbon carbohydrate (see Chapter 2).

The formed carbohydrate becomes the building block for the formation of other organic compounds, some of which form the cellular structure of the plant. These three elements, C, H, and O, constitute from 90–95% of the dry weight of the plant. The remaining 5–10% of the dry weight is made up of the remaining 13 essential elements, as well as other nonessential mineral elements.

MINERAL ELEMENTS

The mineral elements, 13 in number and essential for plants, are taken up in their either cationic or anionic forms (Table 4.3) from the rooting medium through the roots.

TABLE 4.3
The Thirteen Essential Elements and Their Form Taken Up by the Root

Element	Symbol	Ionic Form
	Cations	
Ammonium	NH_4	NH_4^+
Calcium	Ca	Ca^{2+}
Copper	Cu	Cu^{2+}
Iron	Fe	Fe^{2+}, Fe^{3+}
Magnesium	Mg	Mg^{2+}
Manganese	Mn	Mn^{2+}, Mn^{4+}
Potassium	K	K^+
Zinc	Zn	Zn^{2+}
	Anions	
Boron	B	BO_3^{3-}
Chloride	Cl	Cl^-
Molybdenum	Mo	MoO^-
Nitrate	NO_3	NO_3^-
Phosphorus	P	$H_2PO_4^-$, HPO_4^{2-}
Sulfur	S	SO_4^{2-}

Details on these 13 elements, their characteristics, plant and soil chemistry, common chemical forms as fertilizers, concentration and form in hydroponic solutions, and their content in the tomato plant are given in Appendix III by element.

The relationships between and among the elements can have as a significant effect on the plant as the concentration of the element itself, as has been reviewed by Barber (1995), Barber and Bouldin (1984), and Mills and Jones (1996). Elevated elemental levels can create toxicities in the plant, the concept of dose relationships having been reviewed by Berry and Wallace (1981).

ELEMENTAL FORM

The form of an element, whether that found in a nutrient solution or in the solution of a mineral soil or soilless media, can have a marked effect on its utilization and effect on the plant (Barber and Bouldin, 1984; Barber, 1995). The two elements of that form that can affect a tomato plant are nitrogen (N) and iron (Fe).

NITROGEN (N)

The two ionic forms of N utilized by the tomato plant are ammonium (NH_4^+), which is a cation, and nitrate (NO_3^-), which is an anion. Both N forms are found in fertilizers and reagents used to constitute a nutrient solution, and in widely varying concentrations in either the soil solution or water in a soilless media. When NH_4 is the major N source, toxicity can occur resulting in a significant reduction in fruit yield (Barker and Mills, 1980). However, the NH_4^+ form is more readily utilized by the tomato plant in its initial stage of development, which benefits early growth and development.

However, when the tomato plant enters its reproduction stage, NH_4 can be detrimental by affecting both plant growth and fruit yield as well as increasing the incidence of blossom-end rot (BER) in fruit, a phenomenon that has been frequently observed and reported (Pill and Lambeth, 1997; Pill et al., 1978). In a long-term greenhouse tomato project described by Bruce et al. (1980), the change in N form from an NH_4–N-containing fertilizer to an all NO_3–N formulation, almost entirely eliminated the occurrence of BER in fruit as well as plant death from vascular decay.

Hartman et al. (1986) studied the influence of varying ratios of NH_4:NO_3 on tomato plant growth, fruit yield, and incidence of BER using the standing-aerated hydroponic growing technique. One set of their results is shown in Table 4.4.

In the nutrient solution when the NH_4 ratio was greater than 25% of total N, there was a decrease in the number of fruit and a significant decrease in the fresh weight of fruit. In addition, the number of fruit with BER was doubled when NH_4 was present in the nutrient solution, irrespective of its ratio with NO_3, suggesting that any presence of NH_4 can significantly increase the incidence of BER.

In addition to changes in fruit yield, Hartman et al. (1986) also observed significant changes in the elemental composition of the plant itself. With an increasing percentage of NH_4 in the nutrient solution, there was a corresponding increase in the P content of the plant; and a significant decrease in the K, Ca, and Mg contents.

TABLE 4.4
Fruit Yield and Blossom-End Rot (BER) Incidence of 'Floradel' Tomato as Affected by NO_3:NH_4

NO_3:NH_4 (%)	Fruit Number	Mean Fruit Fresh Weight (g)	Number Fruit with BER
100:0	20	1161	6
75:25	20	679	14
50:50	17	413	11
25:75	18	490	12

In the fruit, the P content increased and the K content decreased, while the Ca and Mg contents were not affected.

Adams (1986) reported that adequate N is essential for plant growth, influencing flower numbers and yield of fruit; however, excessive N will significantly increase the percentage of unevenly ripened (blotchy) fruit (Grierson and Kader, 1986).

Iron (Fe)

Maintaining sufficient Fe concentration in a nutrient solution can be difficult when using various inorganic forms of Fe, such as ferrous sulfate ($FeSO_4$) or iron ammonium citrate as recommended in the Hoagland formulation (see Table 6.5). Chelated Fe, normally FeEDTA, is the generally recommended form, a form that remains relatively stable in solution. Other chelated forms of Fe can be used as has been described by Wallace (1983).

CRITICAL MAJOR ELEMENTS

There are three elements, Ca, K, and P, which are particularly critical in the production of tomatoes in soil–field systems, but probably more so for hydroponically grown plants.

Calcium (Ca)

The lack of Ca is intimately tied to the occurrence of BER in fruit. Grierson and Kader (1986) found that when the dry weight Ca level in the fruit was 0.12%, no BER occurred but it did occur when the Ca content was 0.08%. However, the occurrence of this fruit disorder is not just a Ca deficiency problem but is due to a combination of factors that restrict the movement of Ca into the fruit. The most common inducing factor is moisture stress, due to either an excess or a deficiency of water. Under high humidity, or excess water conditions, or both, transpiration is slowed in the plant. Since Ca moves in the plant by means of the transpiration stream, a reduction in water movement within the plant reduces the amount of Ca-carrying water reaching the developing fruit. Under water stress, the same phenomenon occurs and BER occurs in developing fruit.

Any factor that would restrict the uptake of Ca through the roots—low soil pH and imbalance among the major cations, K and Mg plus NH_4—whether in soil or in a nutrient solution, can interfere with Ca uptake. Hartman et al. (1986) observed a significant decrease in the Ca content of the tomato plant with increasing NH_4 in the nutrient solution coupled with an increase in the incidence of BER. Having a sufficient concentration of Ca in the soil or a nutrient solution is important, but frequently it is the balance among the major cations that interferes with the uptake of Ca (Voogt, 1993; Barber, 1995).

Applying a Ca-containing solution to the foliage or developing fruit will have little influence on the Ca content of developing fruit, thereby reducing the incidence of BER.

Phosphorus (P)

For a number of years, P deficiency has been the major concern, its insufficiency reducing or slowing plant growth. In soil, maintaining the soil P level above a "low" soil test level and keeping the soil water pH within the optimum range (6.0–6.5) should be the goal for a fertilizer–lime recommendation. However, keeping the P soil level just under the "high" and certainly under a "very high" test level avoids possible interactions with other elements such as Zn, and thereby creating an imbalance among elements that could result in a nutrient element deficiency (Lindsay, 1979; Barber, 1995).

For a hydroponic solution, 20–50 mg P L^{-1} (ppm) in the nutrient solution is recommended (see Table 6.6), although that level can be toxic to plants if continuously maintained as was discovered by Asher and Loneragan (1963) and Asher and Edwards (1978). However, with most hydroponic systems, the level of P is not continuously maintained (Bugbee, 1995). At what nutrient solution level of P an excess can occur that would affect the tomato plant adversely has not been well defined.

In tomato leaf tissue, Peñalosa et al. (1989) have suggested 0.66% P as the upper limit, while Asher and Loneragan (1963) have suggested 0.80%, with toxicity occurring between 0.90 and 1.80%. Adams (1986) indicated P leaf levels in excess of 1.00% as being toxic. The incidence of high plant P levels observed in summaries of plant analysis results by the author suggests that P excess may be a factor affecting plant growth and fruit yield because a significant number of tomato leaf tissue samples have been found to have P levels in excess of 1.00% (Jones, 1998). The interactions occurring between P and the micronutrients Fe, Mn, and Zn have been studied, showing that increasing P can interfere with the uptake and function of these elements in plants (Lindsay, 1979; Barber, 1995). Therefore, a high P level in the plant may be seen as a deficiency of one of these elements.

One of the factors that needs to be investigated is the influence that a periodic and temporary anaerobic condition occurring in a hydroponic growing media might have on P uptake by the tomato plant. The author believes that when an anaerobic condition exists, P uptake is enhanced, thereby reaching toxic levels in the plant if the anaerobic condition occurs repeatedly over an extended period of time.

Whether growing in soil or in hydroponic culture the level of P should be carefully monitored by periodic analyses of the growing media, the nutrient solution, and the tomato plant. If the P level in the tomato plant exceeds 1.00%, then a significant and immediate change in the supply of P must be made.

Potassium (K)

An inadequate supply of K to the tomato plant will result in uneven ripening of the fruit (Adams, 1986) and will reduce its keeping quality (Gallagher, 1972). Adams (1986) reported that fruit acidity increased linearly from 6.5–8.5 meq 100 mL^{-1} juice as leaf K increased from 3.8–6.0%. For the soil grower, the soil test level for K should be maintained at high, but not over that level since K can significantly interfere with the availability of Mg and Ca (Barber, 1995). Potassium soil availability is significantly affected by soil aeration; and by keeping the soil too wet, K uptake can be reduced.

In the hydroponic growing of tomato, the K level in solution normally ranges between 100–200 mg L^{-1} (ppm) (see Table 6.6). Adams (1986) found that the yield of fruit per plant in hydroponic culture peaked at about 150 mg K L^{-1} in the nutrient solution. Grierson and Kader (1986) observed that as the K content available to the plant increased, the percentage of unevenly ripened (blotchy) fruit decreased. Therefore, it is not uncommon to increase the K content in the nutrient solution when the tomato plant begins to fruit. Plant analyses or tissue tests (discussed later in this chapter) should be used to monitor the tomato plant to ensure that adequate K exists in the plant. Note the change in the deficient K level in the tomato plant leaves during vegetative growth (<1.5%) and that when fruiting (<2.5%), as is shown in the section on essential element levels.

MICRONUTRIENTS

The relative response of the tomato plant to the micronutrients has been given by Lorenz and Maynard (1988) and Vitosh et al. (1994) as follows:

	Relative Response	
Micronutrient	**Lorenz and Maynard**	**Vitosh et al.**
Boron (B)	Medium	Medium
Chlorine (Cl)	—	Not known
Copper (Cu)	Medium	High
Iron (Fe)	High	High
Manganese (Mn)	Medium	Medium
Molybdenum (Mo)	Medium	Medium
Zinc (Zn)	Medium	Medium

Although the micronutrients are essential elements and insufficiencies can occur, they as a group have received less attention than the major elements. Those micronutrients that have been studied for tomato are B, Fe, and Zn, the main focus being on deficiency. Micronutrient deficiencies are not common except on very sandy soils, on high pH soils, or in instances when imbalances occur due to major element excesses such as P. High plant Mn levels can occur when the soil is very acid, or when soils are steam sterilized (Brooks, 1969; Wittwer and Honma, 1969; Ward, 1977).

ESSENTIAL ELEMENT LEVELS

The level of an essential element in the tomato plant determines the plant's nutritional status, which, in turn, affects plant growth as well as fruit yield and quality. What is considered the optimum concentration level for the essential elements in the tomato plant has been fairly well established, the average normal range and deficient level for the essential elements found in tomato plant leaves being:

Element	Normal Range (%)	Deficient (%)
	Major Elements	
Nitrogen (N)	2.8–6.0	<2.0
Phosphorus (P)	0.3–0.9[a]	<0.2
Potassium (K)	2.5–6.0[b]	<1.5 vegetative (<2.5 fruiting)
Calcium (Ca)	0.9–7.2[c]	<1.0
Magnesium (Mg)	0.4–1.3	<0.3
Sulfur (S)	0.3–4.2	—
	[$mg\ kg^{-1}$ (ppm)]	[$mg\ kg^{-1}$ (ppm)]
	Micronutrients	
Boron (B)	25–100	<20
Chlorine (Cl)	Not known	—
Copper (Cu)	5–20	<4
Iron (Fe)	40–300	<40 (<50 may be deficient)
Manganese (Mn)	40–500	<30
Molybdenum (Mo)	0.9–10.0	Not known
Zinc (Zn)	20–100	<16

[a] Levels in excess of 1.00% can be detrimental to the plant.
[b] Relationship between K and Ca may be more important than either element alone.
[c] Levels less than 1.50% may result in significant BER in fruit.

More specific ranges of elemental sufficiency are given by element in Appendix III. The elemental content of the tomato plant can be easily determined by means of a plant analysis, or tissue test, or both, techniques that are discussed in detail later in this chapter.

NUTRIENT ELEMENT UPTAKE PATTERNS WITH TIME

Halbrooks and Wilcox (1980) determined the uptake patterns for the elements N, P, K, Ca, and Mg for field-grown tomato plants 21 days after emergence to harvest.

Dry weight accumulation per plant was linear after 56 days at 5.5 g day^{-1}; and after 70 days, dry weight accumulation was mainly as fruit, accumulating at 4.9 g day^{-1}. Nutrient element accumulation after 70 days was

Element	Milligrams per day accumulation	
	Vines	Fruit
Nitrogen (N)	20	150
Phosphorus (P)	2	21
Potassium (K)	25	231
Calcium (Ca)	120	6
Magnesium (Mg)	10	10

The accumulation of N, P, and K by the fruit was almost 10 times that of the vines, while Ca accumulation by fruit was 1/20th that in the vine, and Mg was the same for vines and fruit. These results clearly show that there must be sufficient supplies of N, P, and K available to the plant when it is fruiting, with the rooting medium plus that already in the plant being the sources of supply.

VISUAL SYMPTOMS OF DEFICIENCY AND EXCESS

When an essential element is below or above the sufficiency range in the plant, visual symptoms usually appear that can be used to identify the deficiency or excess. Roodra van Eysinga and Smilde (1981) have given descriptions of these visual symptoms by element for the tomato plant when grown hydroponically. These descriptions are given in Table 4.5 as well as the incidence that can lead to either deficiency or excess.

TABLE 4.5
Descriptions of Visual Nutrient Element Deficiency Symptoms and Excess Symptoms for Tomato Plants Grown Hydroponically

Nitrogen (N) Deficiency

Symptoms

Shoot growth is restricted and the plant is spindly in appearance. At first, the lower leaves turn yellowish green. In severe deficiency, the entire plant turns pale green. The leaflets are small and erect, and the major veins look purple, especially underneath. Fruits remain small. A nitrogen-deficient crop is susceptible to gray mold (*Botrytis cinerea* Pers. ex. Fr.) and possibly to potato blight [*Phytophthora infestans* (Mont.) de Bary].

Incidence

Nitrogen deficiency can be expected on clay soils in newly built greenhouses if insufficient nitrogen fertilizer is applied and on sandy soils after thorough leaching. It can be induced with straw or with farmyard manure containing much undecomposed straw.

TABLE 4.5 (continued)
Descriptions of Visual Nutrient Element Deficiency Symptoms and Excess Symptoms for Tomato Plants Grown Hydroponically

Phosphorus (P) Deficiency

Symptoms

Shoot growth is restricted and the stem remains thin, but clear symptoms are absent. In more severe deficiency (e.g., in water culture), leaves are small and stiff and curve downward; the upper sides are bluish green and the undersides (including the veins) are purple. The older leaves may turn yellow and develop scattered brownish purple dry spots; they drop prematurely.

Incidence

Phosphorus deficiency may occur on low poorly drained iron-rich (phosphate-fixing) soils. It may also be expected on barren soils landscaped by excavating or pumping in mud or sand, if no phosphate is applied. In old greenhouses, it sometimes occurs after deep digging of the soil. Symptoms of malnutrition may also appear in plants raised on peat substrates if no phosphate is applied. Factors affecting root growth, especially low temperature, promote the deficiency.

Potassium (K) Deficiency

Symptoms

The leaflets of older leaves develop scorched and curled margins and interveinal chlorosis; the smallest veins do not remain green. In some varieties small dry spots with brown margins appear in the chlorotic areas. Plant growth is restricted and the leaves remain small. At a later stage chlorosis and necrosis spread to younger leaves, after which the severely yellowed and curled older leaves drop off. Uneven ripening of the fruit can be expected. A potassium-deficient crop is susceptible to gray mold.

Incidence

Because potassium has a specific influence on fruit quality, it is generally applied liberally and deficiency is rare. It can be expected on potassium-fixing clay, on coarse-textured sand, on peat substrate, and in nutrient solution, if fertilizing is neglected.

Magnesium (Mg) Deficiency

Symptoms

The margins of the leaflets of older leaves show a yellowish discoloration that spreads toward the interveinal tissues. The smallest veins do not remain green. Yellowing gradually spreads from the base to the top of the plant. In the bright yellow to orange leaf tissue, many unsunken necrotic spots often develop and may coalesce into brown bands between the veins. At first, plant habit and leaf size are normal, and the petioles are not curved. As deficiency becomes more severe, the older leaves die and the whole plant turns yellow. The symptoms may vary between varieties and growing conditions. Fruit production is not seriously affected by moderate deficiency; only in severe deficiency is it markedly reduced.

Incidence

Slight magnesium deficiency occurs in almost all greenhouses and with all soil types. More severe deficiency can be expected on coarse-textured sandy soils. It is promoted by low pH and high potassium status of the soil, and by inadequate supply of nitrogen fertilizer.

TABLE 4.5 (continued)
Descriptions of Visual Nutrient Element Deficiency Symptoms and Excess Symptoms for Tomato Plants Grown Hydroponically

Calcium (Ca) Deficiency

Symptoms

At first, the upper sides of the young leaves are dark green, except for the pale margins; the undersides turn purple. The leaflets remain tiny and are deformed and curled up. Leaf tips and margins then wither and the curled petioles die back. The growing point dies. At this stage, interveinal chlorosis and scattered necrotic spots appear in leaflets of older leaves. Those leaves soon die. Fruits show BER. The roots develop poorly and are brownish.

Incidence

In practice, calcium deficiency occurs rarely in vegetative parts. It may arise with plants on oligotrophic peat and in nutrient solution if calcium supply is neglected. BER occurs on acid soils and on soils with a high salt content.

Sulfur (S) Deficiency

Symptoms

At first, plant habit and leaf size are normal. Stem, veins, petioles, and petiolules turn purple; and leaves turn yellow. The leaflets of older leaves show necrosis at tips and margins, and small purple spots appear between the veins. Young leaves are stiff and curl downward. (Such curling does not normally occur in a nitrogen-efficient plant.) Eventually, large irregular necrotic spots appear on those leaves.

Incidence

Sulfur deficiency is unknown in commercial greenhouses. It may occur in crops grown on peat substrates or in nutritional solution if no sulfur fertilizers are used.

Boron (B) Deficiency

Symptoms

In commercial crops the most striking symptom is a yellow to orange discoloration of the leaflets, particularly of the downcurved top leaflets. In severe deficiency, shoot growth is restricted and later the growing point withers and dies. At about the same time, a slight interveinal chlorotic mottling appears in leaflets of younger leaves. These leaves remain small and curl inward and are deformed. The smallest leaflets turn brown and die. Laterals develop, but growth soon stops as growing points die. Stem, petioles, and petiolules are very brittle, causing leaves and leaflets to break off suddenly, except in some tougher cultivars. The veins of the leaflets, and especially the top leaflets, are clogged. This symptom is specific and occurs even with moderate deficiency. Roots grow poorly and turn brown. Fruits may be malformed with brown lesions in the pericarp.

Incidence

Boron deficiency frequently occurs on sandy loam soils. It is aggravated by liming, by using much oligotrophic peat, and by undermanuring. The quality of irrigation water plays an important role.

TABLE 4.5 (continued)
Descriptions of Visual Nutrient Element Deficiency Symptoms and Excess Symptoms for Tomato Plants Grown Hydroponically

Copper (Cu) Deficiency

Symptoms
Stem growth is somewhat stunted. The leaves are bluish green. The margins of leaflets of middle and of younger leaves curl into a tube toward the midribs. Chlorosis and necrosis are absent. Terminal leaves are very small, stiff, and contorted. Petiolules bend typically downward, directing opposite tubular leaflets toward each other. Later, necrotic spots develop alongside and on the midribs and the larger lateral veins.

Incidence
Copper deficiency rarely occurs in commercial greenhouses. The disorder may occur in plants on oligotrophic peat or nutrient solution.

Iron (Fe) Deficiency

Symptoms
Chlorosis develops in the terminal leaves. At first even the smallest veins remain green to produce a fine reticular pattern of green veins on yellow leaf tissues. Later, chlorosis increases in intensity and extends to the smaller veins. Eventually, affected leaves turn completely pale yellow or almost white; necrosis is not severe. Symptoms progress from the terminal to the older leaves. Growth is stunted and newly formed leaves remain small.

Incidence
Iron deficiency may be expected on fine sands or calcareous loam soils of weak structure. It occurs rather patchily, mainly in plants bearing a heavy crop. It may also occur in plants on peat substrate or in nutrient solution without iron.

Manganese (Mn) Deficiency

Symptoms
Middle and older leaves, and later the younger leaves, turn pale. This produces a characteristic checkered pattern of green veins and yellowish interveinal areas. Later, small gradually expanding necrotic spots appear in the pale areas, especially near the midribs. Chlorosis is less severe than in iron-deficient plants (where entire leaves may turn yellowish white). Another difference is that chlorosis is not confined to younger leaves.

Incidence
Manganese deficiency occurs on calcareous loam soils and on overlimed sand and peat soils. With nutrient solution, manganese deficiency may occur if manganese is not applied.

Molybdenum (Mo) Deficiency

Symptoms
The leaflets show a pale to yellowish interveinal mottling. The margins curl upward to form a spout. The smallest veins turn yellow. Necrosis starts in the yellow areas, at the margins of the top leaflets, and finally includes entire composite leaves that shrivel. Symptoms progress from the older to the younger leaves, but the cotyledons stay green for a long time.

TABLE 4.5 (continued)
Descriptions of Visual Nutrient Element Deficiency Symptoms and Excess Symptoms for Tomato Plants Grown Hydroponically

Incidence

Molybdenum deficiency may occur on acid soils low in phosphate and rich in iron. It is common in plants grown in oligotrophic peat without supplementary molybdenum.

Zinc (Zn) Deficiency

Symptoms

Terminal leaves remain small and the leaflets show slight discoloration between the veins. Growth is stunted. Older leaves are also smaller than normal. There is little chlorosis in those leaves but irregular shriveled brown spots develop, especially on the petiolules, but also on and between the veins of the leaflets. Petioles curl downward, and complete leaves coil up. Necrosis progresses rapidly; within a few days the entire foliage may wither.

Incidence

Zinc deficiency does not occur in greenhouse crops, unless grown in nutrient solution without zinc.

Excess Nitrogen (N)

Symptoms

Growth is restricted. Leaves are shorter than normal, look stiff, and are deep green. In acute toxicity, leaves lose turgor, margins desiccate, and sunken watery spots appear, with the affected leaf tissue dying back and turning whitish gray.

Incidence

Excess nitrogen is induced by heavy dressings of nitrogen or of organic materials like dried blood. Acute toxicity is caused by excessive topdressing with inorganic nitrogen or by uneven distribution.

Excess Manganese (Mn)

Symptoms

The plant is somewhat spindly and restricted in growth. Terminal leaves remain tiny and the leaflets show interveinal chlorosis. Many necrotic interveinal spots develop in the leaflets of older leaves, making them look dirty. Later the midrib and the larger lateral veins die. The leaves are then shed, the older ones first.

Incidence

Disinfection of soils by steaming releases much plant-available manganese that may induce manganese toxicity. Low pH promotes the disorder.

Excess Zinc (Zn)

Symptoms

The plant is spindly and growth is severely stunted. Symptoms differ from those produced by iron deficiency in that the younger leaves are extremely small, the leaflets show interveinal chlorosis, and the undersides turn purple. Older leaves are strongly downcurved. Purplish tints develop on the undersides, spreading form the margins inward. Leaves may later turn yellow (reddish brown veins excepted) before dropping.

TABLE 4.5 (continued)
Descriptions of Visual Nutrient Element Deficiency Symptoms and Excess Symptoms for Tomato Plants Grown Hydroponically

Incidence

Zinc toxicity may occur in greenhouses if condensation drips from the galvanized frame onto the plants. Care should be taken with galvanized materials. The watering systems should not contain any such material.

Excess Boron (B)

Symptoms

Tips and margins of the leaflets of older leaves and of cotyledons become scorched and curl. Later, sunken desiccating spots may develop, sometimes surrounded by brown concentric rings. The downcurved leaflets feel dry and papery, and finally drop. Symptoms progress from older to younger leaves. At first, the plant top looks almost normal, but later terminal leaves curl too.

Incidence

Boron toxicity is easily caused by excess boron fertilizer. Special care should be taken in applying such materials.

Source: Roorda van Eysinga, J.P.N.L. and K.W. Smilde. 1981. *Nutritional Disorders in Glasshouse Tomatoes, Cucumbers, and Lettuce.* Centre for Agriculture Publishing and Documentation, Wageningen, The Netherlands.

A quick description of elemental deficiency and toxicity symptoms occurring in tomato plants when hydroponically grown has also been categorized by Taylor (1983), the symptoms and associated elements being listed in Table 4.6.

Photographs of visual deficiency or excess (toxicity) symptoms of the essential elements in tomato are scattered among various publications (Roorda van Eysinga and Smilde, 1981; Bergmann, 1992; Bennett, 1993; Weir and Cresswell, 1993; Anon., 1998a). There is also a video form that discusses tomato plant nutrition with pictures of typical insufficiency symptoms (Jones, 1993b). In all these picture publications, the photographs too frequently are of relatively poor quality, or of only single leaves, only of very young plants; or include only some of the essential elements. Unfortunately, there are very few photographs of toxicity symptoms due to an excess or imbalance of the essential elements. Techniques for diagnosing mineral disorders in plants for greenhouse crops are discussed in the book by Winsor and Adams (1987).

Photographs of good quality have appeared on the Internet, which may be the best means of obtaining visual identification of insufficiencies associated with the essential elements. Details on the Internet are given in Chapter 1 of this book.

BENEFICIAL ELEMENTS

Although no such category has been officially established, many believe that more than the 16 essential elements must be present in the plant to ensure maximum

TABLE 4.6
Deficiencies and Toxicities of Mineral Elements in Hydroponic Nutrient Solutions

Symptoms	Primary Responsible Elements								
	N	P	K	Ca	Mg	Fe	Zn/Cu	B	Mo
Deficiencies									
General stunted, spindly or restricted growth	X	X			X				
Chlorosis (yellowing)									
Older leaves	X		X		X				X
Younger leaves						X	X	X	
Necrosis (drying or scorched leaves)			X		X	X	X		
Buds inhibited or die				X			X		
Root tips die				X				X	
Leaf tip burn old leaves	X								
Leaf curl			X	X	X	X			
Light green color	X								
Dark green and purple		X							
Mottling (blotchy color)			X		X		X		
Tips and margins cupped up					X				
Leaf growth stunted							X		
Hard stems			X					X	
Poor root systems			X		X				
Soft stems	X		X						
Old leaves drop		X							
Blossom drop						X			
Toxicities									
Lush dark green									
foliage at first	X								
Soft elongated stems	X								
Restricted flowering and fruit	X								
Large light green leaves					X				
Restricted roots	X								
Chlorosis							X	X	
Leaf curl					X				
Necrosis and later plant death								X	
Results in deficiencies or other elements				X		X			

Source: Taylor, J.D. 1983. *Grow More Nutritious Vegetables without Soil*. Parkside Press Publishing, Santa Ana, CA.

growth. In earlier times in the hydroponic culture of plants, an A–Z Solution containing 20 elements (see Table 4.7) was added to the nutrient solution containing the known essential elements. A detailed description of the composition of the A–Z Solution and of its use has been described by Jones (1997b). The purpose of the A–Z Solution was to ensure that almost every element found in the soil would be included in the nutrient solution. The A–Z Solution is not used today, but known plant responses to many of the elements in the A–Z Solution as well as others on the growth of plants have been observed; and a review of these observations has been made by Kabata-Pendias and Pendias (1994) and Pais and Jones (1997).

TABLE 4.7
Elements in the A–Z Solution

Element (Symbol)	Element (Symbol)
Aluminum (Al)	Lithium (Li)
Arsenic (As)	Lead (Pb)
Barium (Ba)	Mercury (Hg)
Bismuth (Bi)	Nickel (Ni)
Bromine (Br)	Rubidium (Rb)
Cadmium (Cd)	Selenium (Se)
Chromium (Cr)	Strontium (Sr)
Cobalt (Co)	Tin (Sn)
Fluorine (F)	Titanium (Ti)
Iodine (I)	Vanadium (V)

There are two elements—silicon (Si) and nickel (Ni) (Eskew et al., 1984; Brown et al., 1987)—that have been suggested as being essential for plants. The major role of Si has been found to be in the strengthening of the stem of rice (Takahnashi et al., 1990), as well as other grain crops, plus the possibility that Si may also contribute to the stem strength of the tomato plant. Silicon has also been found to be a factor in preventing the penetration of fungus hypha (disease resistance) into plant leaf cells (Bélanger et al., 1995), therefore making the plant more resistant to fungus attack. Since this disease resistance aspect of Si could be of major benefit for tomato plants being grown hydroponically in the greenhouse, several soluble formulations of Si are available for addition to the nutrient solution. Also, Epstein (1994) found that Si enhances the tolerance of plants to elevated concentrations of both aluminum (Al) and iron (Fe).

There are three other elements, sodium (Na), vanadium (V), and cobalt (Co), which fall into the category as being beneficial, because Na can partially substitute for potassium (K) and V for molybdenum (Mo), while Co is required by the nitrogen (N_2)-fixing bacteria in leguminous plants. However, none of these three elements have been found to be beneficial to the tomato plant, except for Na that might be a factor in enhancing the flavor of tomato fruit (Gough and Hobson, 1990).

PLANT NUTRIENT ELEMENT MANAGEMENT

The success of any growing system is based on the ability of the grower to maintain the nutrient element status of the plant without incurring insufficiencies; this is not an easy task, particularly when growing in soil in the field. Even successfully growing plants hydroponically in the greenhouse can be a formidable task. Procedures that would apply to field soil-grown plants have been reviewed by Stevens (1986),

Hochmuth (1996a), and Snyder (1996c), and for drip-irrigated plasticulture systems by Hartz and Hochmuth (1996). For greenhouse-grown plants, Hochmuth (1996b) and Snyder (1997a) provided management information required to maintain plant nutrient element sufficiency.

In an interesting study, Halbrooks and Wilcox (1980) monitored the tomato plant from seedling stage to full fruit production for the major element requirements, demonstrating the need for a continuous supply of N, K, and K for the maximum fruit yield to be obtained.

PLANT ANALYSIS

The nutrient element status of a plant is best determined by means of a plant analysis. Unfortunately, the plant part to be sampled for analysis has not been standardized, which can make the interpretation of a plant analysis result difficult. The types of tissue recommended for sampling are

- Compound leaves adjacent to top inflorescences
- Mature leaves from new growth (greenhouse plants)
- Mature leaf (healthy plant)
- Young mature leaf
- Youngest open leaf blade

It is important to understand that the type of tissue sampled and the time of its sampling (plant stage of development) are correlated with a particular set of interpretation values; therefore, considerable care needs to be exercised when interpreting a plant analysis result to ensure that the interpretation data used relate to the plant part assayed and to the time of sampling and plant stage of growth. To illustrate what elemental ranges for sufficiency exist based on stage of plant development for a particular plant part, the interpretative ranges given by Reuter and Robinson (1997) are shown in Table 4.8.

Various sources of plant analysis interpretative data, grouped by plant part assayed and time of sampling, are given by element in Appendix III.

It is important to make contact with the laboratory prior to collecting a tissue sample to ensure that the proper sample is taken at the specified time since interpretation of the assay result is correlated with these sampling parameters, as has been described by Morard and Kerhoas (1984), Mills and Jones (1996), and Reuter and Robinson (1997).

A book on the plant analysis technique has been edited by Kalra (1997), and a video on the plant analysis technique by Jones (1993c) is available.

SAMPLING INSTRUCTIONS

The author recommends that the very end leaf on the compound leaf (Figure 4.1) be taken from a recently mature leaf that would normally be at that level of the plant where the most recent fruit cluster is developing. Select 30 to 50 plants for sampling. Sampling recommendations are as follows:

TABLE 4.8
Summarized Nutrient Element Levels for Interpreting a Tomato Leaf Analysis

	Stage of Development[a]			
Element	**Early Flower** (%)	**Early Fruit Set** (%)	**First Mature Fruit** (%)	**Midharvest** (%)
Nitrogen (N)				
Critical	4.90	—	4.45	—
Adequate	5.0–6.0	4.6–6.0	4.5–4.6	4.5–5.5
Phosphorus (P)				
Adequate	0.4–0.9	0.3–0.9	0.4–0.9	0.6–0.8
Potassium (K)				
Adequate	3.8–6.0	3.3–5.0	3.0–5.0	3.4–5.2
Calcium (Ca)				
Adequate	1.5–2.5	1.4–3.2	1.4–4.0	2.0–4.3
Magnesium (Mg)				
Adequate	0.4–0.6	0.39–0.71	0.4–1.2	0.51–1.30
Sodium (Na)				
Adequate	0.1–0.4	—	—	—
Chloride (Cl)				
Adequate	0.5–2.5	—	—	—
	[mg kg^{-1} (ppm)]	**[mg kg^{-1} (ppm)]**	**[mg kg^{-1} (ppm)]**	**[mg kg^{-1} (ppm)]**
Nitrate–N (NO_3–N)				
Critical	760	760	1,120	—
Adequate	1,100–1,240	1,000–1,200	1,790	1,600
Boron (B)	30–100	—	—	—
Copper (Cu)	5–15	—	—	—
Iron (Fe)	60–300	—	—	—
Manganese (Mn)	50–250	—	—	—
Zinc (Zn)	30–100	—	—	—
Molybdenum (Mo)	0.6	—	—	—
Aluminum (Al)	<200	—	—	—

[a] Youngest open leaf blade.

Source: Reuter, D.J. and J.B. Robinson. 1997. *Plant Analysis: An Interpretation Manual,* 2nd ed., CSIRO Publishing, Collingwood, Australia.

- For plants showing signs of stress, collect two sets of leaf material, one set from plants that seem "normal" in appearance and another from those plants showing stress.
- Do not include dead plant material, or select from plants that are disease infested or insect damaged or from plants that have been mechanically damaged.

FIGURE 4.1 Whole tomato leaf showing the detached (top) leaflet to be collected for a plant analysis.

Decontamination

If plants have been sprayed with chemicals, or have been exposed to dusty air, those substances must be removed at the time the tissue is collected by carefully washing the tissue in a 1% P-free detergent solution for about 30 sec, and then rinsed in a flow of clean water. The tissue is then blotted dry and placed in a clean bag.

Drying and Shipping

Fresh tissue should never be placed in a closed plastic bag unless it is kept cool during shipment to the laboratory. Otherwise, allow the tissue to partially dry by leaving it open to clean dry air for at least 24 h, and then place in a clean paper bag for shipment.

Interpretation

An interpretation of a plant analysis should be made by an experienced professional. When a diagnosis is being made for a plant under suspected nutrient element stress, a sample of the rooting medium, whether it be a soil or a hydroponic nutrient solution, should also be collected and assayed along with the plant tissue. All the parameters associated with the production of the tomato plant need to be provided to the interpreter. Normally a questionnaire is provided by the laboratory providing the plant analysis service, which asks for information about the crop being sampled.

A recommended system of diagnosis is presented in a Potash and Potash Institute publication (Anon., 1991). A complete crop diagnostician needs to investigate all aspects which that include investigation of:

- Root zone
- Temperature
- Soil pH
- Insects
- Diseases
- Moisture conditions
- Soil salinity problems
- Weed identification
- Herbicide damage
- Tillage practices
- Hydrid or variety
- Plant spacing
- Water management
- Date of planting
- Fertilizer placement

LABORATORY SERVICES

A list of plant analysis laboratories in the United States and Canada can be found in two publications, one prepared by Downing and Associates (Downing, 1997) and the other published by CRC Press (Anon., 1998b).

TISSUE TESTING

The nitrate–nitrogen (NO_3–N), P, and K status of the tomato plant can be determined using field tissue test kits. The leaf petiole is the tissue normally assayed. Coltman

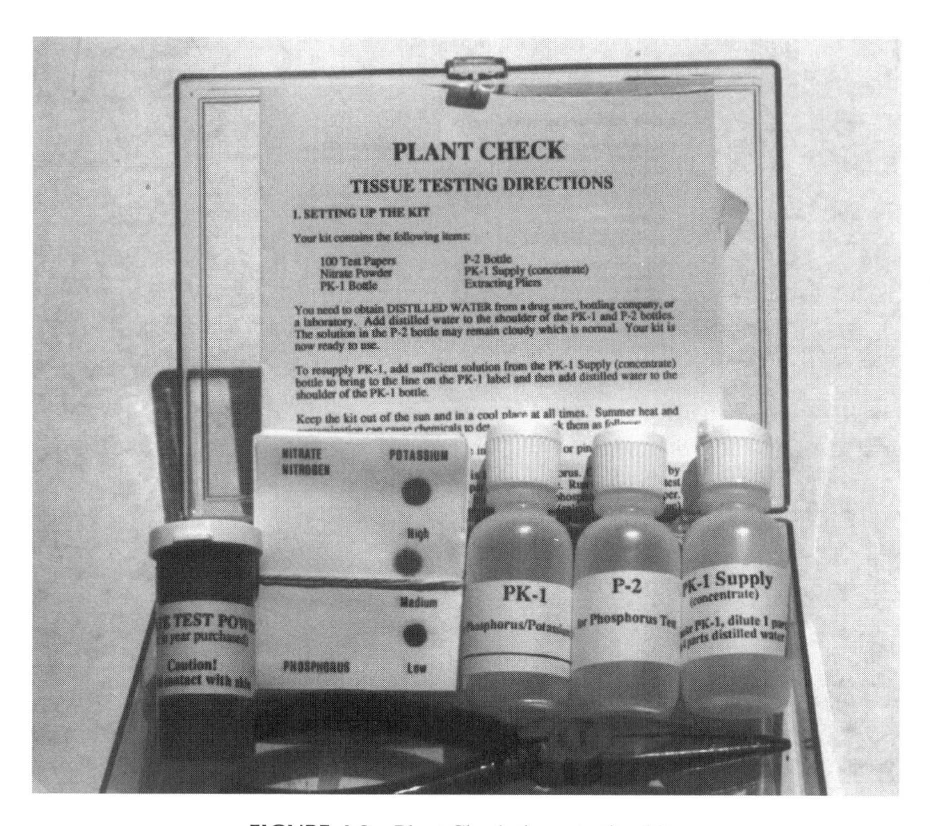

FIGURE 4.2 Plant Check tissue testing kit.

(1987, 1988) has described a NO_3–N petiole testing procedure for regulating the N fertilization of field and greenhouse tomatoes, respectively. Interpretative data for petiole NO_3–N content have been given by Hartz and Hochmuth (1996) and Reuter and Robinson (1997); and for petiole K content, by Hartz and Hochmuth (1996). Beverly (1994) has described a stem tissue testing procedure for NO_3 and K evaluation for tomato seedlings. Tissue testing procedures have been described by Syltie et al. (1972) and Jones (1997a). A video showing how tissue tests are conducted is available (Jones, 1993d). Tissue testing interpretative data are given in Appendix III.

Tissue testing kits can be obtained from:

- HACH Chemical Company, P.O. Box 389, Loveland, CO 80539
- Plant Check kit (Figure 4.2) from Spectrum Technologies, Inc., 23839 West Andrew Road, Plainfield, IL 60544

CONCENTRATION UNITS

Those who are familiar with the scientific literature find that there has not been a standardization of units for reporting a plant analysis result. However, in most of the popular literature, percentage (%) is the concentration unit used for the major elements, and parts per million (ppm) or milligrams per kilogram (mg kg^{-1}) are used for the micronutrients. Concentration units are normally based on the dry weight of tissue unless otherwise designated as being either on a wet weight or a unit of weight basis.

A comparison of commonly used concentration units for the major elements and micronutrients in plant tissue is as follows (concentration levels were selected for illustrative purposes only):

Element	Unit			
Major elements	**%**	**g kg^{-1}**	**cmol(p+) kg^{-1}**	**cmol kg^{-1}**
Phosphorus (P)	0.32	3.2	—	10
Potassium (K)	1.95	19.5	50	50
Calcium (Ca)	2.00	20.0	25	50
Magnesium (Mg)	0.48	48.0	10	20
Sulfur (S)	0.32	3.2	—	10
Micronutrients	**ppm**	**mg kg^{-1}**	**mmol kg^{-1}**	
Boron (B)	20	20	1.85	
Chlorine (Cl)	100	100	2.82	
Iron (Fe)	111	111	1.98	
Manganese (Mn)	55	55	1.00	
Molybdenum (Mo)	1	1	0.01	
Zinc (Zn)	33	33	0.50	

5 Field Production in Soil

CONTENTS

Geisenberg and Stewart (1986) have written a comprehensive chapter on field crop management procedures, discussing timing of operations, land preparation, seedbed conditions, fertilizer and irrigation requirements, crop spacing, transplant setting, plant training, fruit ripening, harvesting, and fruit handling. Stevens (1986) discusses the future of field crop production with the primary requirements based on the improvement in crop tolerances toward heat, cold, drought, and soluble salts as well as increased resistance to diseases and insects. Quality improvements based on increased total solids content of fruit plus color and flavor are future goals. Longer shelf life for fresh market fruit is another goal. All these improvements are and will continue to come from breeding efforts and by bringing genetic engineering techniques into use. New tomato varieties for Florida, for example, are introduced each year, and Maynard (1997) has published lists of current recommended varieties.

The tomato plant grows quite well under a range of soil conditions, the limiting factors being more frequently climatic than soil related. The physical and chemical characteristics of soils most desirable for best plant growth and fertilizer recommendations are discussed in this chapter.

SOIL PHYSICAL CHARACTERISTICS

The tomato plant will grow well under a range of soil textural properties from sandy to fine-textured clays if the soil is well drained (tomato roots will not tolerate waterlogging), has good structure, and is well aerated. Since the tomato plant does not have a single primary taproot but has a branching tap–fibrous root structure, aeration and loose soil conditions are essential for vigorous plant growth and high yield production. The roots will not penetrate compaction pans. Roots of the tomato plant will generally occupy the plow zone or the upper 23 in. (60 cm) of soil, however, with about 70% of the roots being in the top 7.8 in. (20 cm) of soil. On sandy soils, control of the water and nutrient environment can be obtained by some form of plasticulture (Lamont, 1996).

SUBSOIL CHARACTERISTICS

The subsoil may be a factor affecting tomato plant growth, depending on its physical and chemical properties. Root penetration into the subsoil can be blocked by physical factors, such as the presence of a naturally occurring hard pan or produced plow pan, anaerobic conditions, or high moisture content; or by chemical factors, such as low pH (<5.5) and soil fertility, and high salinity [2.5 decisiemens (dS) m^{-1}]. If subsoil conditions are suitable for root penetration, the subsoil can be a source of water and plant nutrients that can make the tomato plant less sensitive to environmental stresses, particularly moisture stress. Depending on what subsoil conditions exist, steps to modify the subsoil to make it more habitable for roots could be of significant benefit.

SOIL FERTILITY REQUIREMENTS

There are specific soil fertility parameters that may be varied depending on the purpose of the fruit, whether for processing or for fresh market, which are required for maximum plant growth and high quality fruit production. In general, the tomato plant grows best on fertile soils, soils that test in the "medium" to "high" level for phosphorus (P), potassium (K), calcium (Ca), and magnesium (Mg). The tomato plant is classified as a "heavy feeder," having high requirements for the elements K, Ca, and iron (Fe); and moderate requirements for nitrogen (N), magnesium (Mg), P, sulfur (S), boron (B), copper (Cu), manganese (Mn), and zinc (Zn).

To produce fruit with high soluble solids content, both K and Ca must be in good supply for plant uptake and utilization. When soils are "low" in Ca and the tomato plant is under moisture stress, blossom-end rot (BER) is likely to occur in the fruit. During initial plant growth, "moderate" to "high" soil test P levels are required, but when the soil tests "very high" in P, and when coupled with high soil pH (>7.0), Zn deficiency is likely to occur (Lindsay, 1979; Barber, 1995).

SOIL PH

The tomato plant grows well within a soil water pH range from 5.5–6.8, with the optimum range being between 6.0 and 6.5. If the soil water pH is below 5.5, Mg availability declines sharply, while the level of available aluminum (Al) and Mn begins to increase significantly; the extent of these changes in availability will vary to some degree on soil type and organic matter content. For many soils, a low pH is coupled with low soil Ca; and when the tomato plant is under stress, BER is likely to occur in the fruit.

When the soil water pH is above 6.8, the availability of Zn, Mn, and Fe will begin to decline, the degree of this pH effect being moderated to some degree by soil type and level of organic matter. However, there is a significant likelihood of deficiency for one or all three of these elements when the soil water pH is greater than 7.0 (Lindsay, 1979; Barber, 1995).

SOIL SALINITY

The tomato plant is considered moderately sensitive to salinity, the maximum soil electrical conductivity (salinity) level without a significant yield loss being 2.5 dS m^{-1}. The relationship between percentage of yield loss and salinity is shown in Table 5.1.

TABLE 5.1
Relationship between Electrical Conductivity and Percentage of Yield Loss

Electrical Conductivity (dS m^{-1})	Yield Loss (%)
1.7	0
2.3	10
3.4	25
5.0	50

Source: Lorenz, O.A. and D.M. Maynard. 1988. *Knott's Handbook for Vegetable Growers*. 3rd ed., John Wiley & Sons, New York.

Soil salinity is becoming a serious problem in several major tomato production areas around the world due to the increasing salinity of applied irrigation water. This is a particular problem in California, the leading state in producing processing tomatoes and second in fresh market tomatoes (see Table 1.2). Significant efforts are now underway to breed varieties that have tolerance to increased soil salinity conditions.

IRRIGATION WATER QUALITY

The quality of irrigation water can be a significant factor, affecting both the tomato plant and the long-term composition of the soil or other rooting media. Both organic and inorganic substances in irrigation water need to be known to assess the impact its use will have on tomato plant culture.

The presence of pesticides, particularly herbicides, in both surface and groundwater supplies is becoming of major environmental concern affecting drinking water quality, water that is also frequently used for irrigation purposes. Most of the commonly used herbicides, atrazine, metachlor, prometon, and simazine, currently being found in both surface and groundwater supplies can be detrimental to the tomato plant. The United States Geological Survey's (USGS) National Water Quality Assessment program, which began in 1991, publishes its latest findings on the its web site at <http://water,wrusgs.gov/pnsp>. Although the presence of these organics is more commonly found in waters in agricultural areas, some urban water supplies can also contain some of these same substances at similar or lower concentrations. Tomato is not tolerant to most herbicides even at the lowest of levels.

The inorganic composition of irrigation water can be a major factor influencing its effect on plants, as has been defined by Farmhand et al. (1985) who have given guidelines for irrigation water quality characteristics in terms of *degree of problem* as:

	Degree of Problem		
Characteristics	**None**	**Increasing**	**Severe**
EC, dS m^{-1}	<0.75	0.75–3.0	>3.0
TDS, mg L^{-1}	<480	480–1920	>1920
Sodium (Na), SAR value	<3	3–9	>9
Chloride (Cl), mg L^{-1}	<70	70–345	>345
Boron (B), mg L^{-1}	1.0	1.0–2.0	2.0–10.0
Ammonium (NH_4) and nitrate (NO_3), mg L^{-1}	<5	5–30	>30
Bicarbonate (HCO_3), mg L^{-1}	<40	40–520	>520

Note: EC, electrical conductivity; TDS, total dissolved solids; SAR, sodium activity ratio.

Many water supplies contain sizable quantities of two essential plant elements, Ca and Mg, which can be initially beneficial to plants. However, the long-term use of high Ca and Mg content water will result in an increase in the pH of the soil (or whatever the rooting medium might be), with the end result being an alkaline condition. This in turn reduces the availability of other essential elements, particularly the micronutrients, Cu, Fe, Mn, and Zn, as well as P (Lindsay, 1979; Barber, 1995).

A determination of both the organic and inorganic composition of irrigation water is highly recommended so that its use can be regulated to minimize any adverse effects, or if found unsuitable, to seek another water source. Depending on what substances are in the water and their concentration or if relatively small quantities of water are needed on a continuing basis, treatment may be economically feasible by removing those substances that would be detrimental to tomato plant growth (Anon., 1997a).

FERTILIZER RECOMMENDATIONS

MAJOR ELEMENTS

In the commercial production of tomato fruit, the fertilizer rate for P and K is normally based on a soil test recommendation, while the N fertilizer rate is determined by crop requirement. The fertilizer recommendation may vary based on use of the fruit, whether for processing or the fresh market, plant density, and whether the plants are being irrigated or must be dependent on natural rainfall. General fertilizer recommendations given by region and states within the United States are listed in Table 5.2.

TABLE 5.2
General Fertilizer Recommendations for N, P_2O_5, and K_2O by Region and State

State/Region	Pounds per Acre (lb A^{-1})		
	N	P_2O_5	K_2O
California	121	80	55
Florida[a]	220	160	300
Florida[b]	175	150	225
Georgia[c]	100–130	60–90	60-90
Georgia[d]	100–160	100–150	100–150
Indiana	50–120	100	240
New York	100–180	50–160	50–160
Wisconsin	100	100	150
New England	130–170	120–150	120–150
Mid-Atlantic[e]	80	200	300
Mid-Atlantic[f]	80	100	100

[a] Mineral soils, low in P and K, irrigated.
[b] Mineral soils, low in P and K, irrgiated, mulched.
[c] Field tomatoes.
[d] Staked tomatoes.
[e] Low soil test P and K.
[f] High soil test P and K.

Phosphorus and K soil test levels and soil type as well as the purpose for production, whether fresh market or processing, will significantly influence a fertilizer recommendation as can be seen in the rates given by Rutgers University (Peet, 1996d) shown in Table 5.3.

Tomato fertilizer recommendations, such as those given for Florida conditions (Hochmuth, 1997) and California (Tyler and Lorenz, 1991), for example, can be obtained from the land-grant colleges and universities through their cooperative extension services. Frequently, this information can also be obtained locally from county agent offices. It is important that specific recommendations be obtained for the soils and climatic conditions of the region rather than relying on general recommendations that may require adjustment to suit local conditions. In addition, crop consultants are able to provide similar assistance to growers through local and national crop consultant organizations.

Based on data obtained from international sources obtained by Halliday and Trenkel (1992), 40–50 ton ha^{-1} tomato crop would require a basic application of 50 kg N ha^{-1}, 150–200 kg P_2O_5 ha^{-1}, and 200–250 kg K_2O ha^{-1}, followed by a topdressing of 100–150 kg N ha^{-1}. From the same authors, a summary of the general fertilizer recommendations given in the *IFA World Fertilizer Use Manual* for several countries are listed in Table 5.4.

von Uexkull (1979) has listed fertilizer recommendations for tomato in several tropical regions of the world; typical fertilizer rates range from 40–120 kg N ha^{-1}, 30–90 kg P_2O_5 ha^{-1}, and 30–90 kg K_2O ha^{-1}. Recommended soil and plant levels are

	Soil		**Plant**	
Element	**Desirable Range (ppm)**	**Toxic Level (ppm)**	**Desirable Range (ppm)**	**Toxic Level (ppm)**
Phosphorus (P)	60–70	—	0.4%	—
Potassium (K)	600–700	—	6.0%	—
Magnesium (Mg)	350–700	—	0.5%	—
Calcium (Ca)	1000	—	1.25%	—
Nitrogen (N)	—	—	3.0–5.0%	—
Boron (B)	1.5–2.5	3.0	40–60 ppm	100
Manganese (Mn)	5–20	80	30 ppm	1000
pH	6.5–7.5			

The source of N, whether ammonium (NH_4)–N or nitrate (NO_3)–N, can significantly affect plant growth if the ratio of NH_4–N:NO_3–N exceeds 1:4 (Barker and Mills, 1980; Hartman et al., 1986). The tomato plant when under high light intensity is affected to a greater degree by NH_4–N nutrition than when under low light intensities (Magalhaes and Wilcox, 1984). High NH_4–N in the soil, occurring from an application of an ammoniacal fertilizer or levels of NH_4–N in a hydroponic nutrient solution exceeding 30 mg L^{-1} (ppm), can result in NH_4 toxicity; the symptoms are leaf cupping, a breakdown of the vascular tissue at the base of the plant, and a high incidence of BER in the fruit. Ammonium–N supplied during initial plant development will not adversely affect plant growth, but high NH_4–N availability after initial fruit set will produce the adverse effects of NH_4 toxicity, particularly by increasing the incidence of BER.

TABLE 5.3
Tomato Fertilizer Recommendations Based on Soil Test Level for Fresh Market and Processing Tomatoes

	N (lbs A^{-1})	Soil Test Phosphorus Level (lb P_2O_5 A^{-1})				Soil Test Potassium Level (lb K_2O A^{-1})			
		Low	Medium	High	Very High	Low	Medium	High	Very High
Fresh Market									
Sandy Loams and Loamy Sands									
Total recommendation	80	200	150	100	50	300	200	100	50
Broadcast and plow down	0	150	100	50	0	250	150	50	0
Broadcast at first cultivation	50	50	50	50	50	50	50	50	50
Side-dress when first fruits are set	30	0	0	0	0	0	0	0	0
Loams and Silt Loams									
Total recommendation	50–80	200	150	100	50	250	150	100	50
Broadcast and plow down or drill deep	50	200	150	100	50	250	150	100	50
Side-dress at first fruit set if needed	25–30	0	0	0	0	0	0	0	0
Processing									
Sandy Loams and Loamy Sands									
Total recommendation	130	200	150	100	50	300	200	100	50
Broadcast and disk in or drill deep	50	200	100	50	0	200	150	50	0
Broadcast at first cultivation	50	50	50	50	50	100	50	50	50
Side-dress when first fruits 1 in. diameter	30	0	0	0	0	0	0	0	0
Loams and Silt Loams									
Broadcast and plow down or drill deep	100–125	250	150	100	50	300	200	100	50

TABLE 5.3
Tomato Fertilizer Recommendations Based on Soil Test Level for Fresh Market and Processing Tomatoes

		Soil Test Phosphorus Level (lb P_2O_5 A^{-1})				Soil Test Potassium Level (lb K_2O A^{-1})			
	N (lbs A^{-1})	Low	Medium	High	Very High	Low	Medium	High	Very High
Side-dress at first fruit set if needed	25–50	200	150	100	50	250	150	100	50

TABLE 5.4
General Fertilizer Recommendations for Tomato in Several Countries

	Kilograms per Hectare (kg ha^{-1})		
Country	**N**	**P_2O_5**	**K_2O**
Senegal (Camberene)	70	200	240
Philippines	96	192	96
Pakistan	150	100	50
Venezuela[a]	96	192	96

[a] After planting: N, 1.8–2.4 kg ha^{-1}; P_2O_5, 3.6–4.8 kg ha^{-1}; K_2O, 1.8–2.4 kg ha^{-1}.

Source: Halliday, D.J. and M.E. Trenkel (Eds.) 1992. *IFA World Fertilizer Use Manual*, pp. 289–290, 331–337. International Fertilizer Industry Association, Paris, France.

In a long-term soilless medium (pine bark) production system described by Bruce et al. (1980), the authors essentially eliminated the occurrence of BER in fruit and the loss of the mature fruiting plants due to vascular decay by substituting an all NO_3–N fertilizer for a NH_4–N-containing fertilizer that was being periodically added to the growing medium.

The level of soil P in many cropland soils has increased substantially as a result of the continued and heavy use of P-containing fertilizers. What effect, if any, this increasing level of soil P is having or will have on tomato plant growth and development has not been adequately explored. The soil should be tested prior to applying a P-containing fertilizer to avoid a P application when not needed, and which would continue to add to an increasing soil P level. Growers should also monitor their tomato plants by means of a plant analysis or tissue tests to ensure that all the elements, including P, are within the desired sufficiency range (see Chapter 4). Although the plant analysis result may not be timely in terms of avoiding an insufficiency for the sampled crop, the assay results can then be used to guide future fertilizer applications so that insufficiencies are not likely to reoccur.

Although Ca, Mg, and sulfur (S) are also essential major elements, little has been published concerning their insufficiencies in soil systems for tomato production. It is assumed that if the soil water pH is kept within the desired range of 6.0–6.5, sufficient Ca and Mg will be present to meet the crop requirement, assuming that dolomitic limestone (contains Mg) is the lime source used to correct soil acidity. However, such verification should be determined by means of a soil test, since a soil test result normally contains an evaluation of the Ca and Mg status of the soil and corrective treatments can be applied, when needed, prior to planting. Since S is found in a number of commonly used fertilizers and S deposition from atmospheric pollution can be quite high (up to 30 lb S A^{-1} $year^{-1}$), its deficiency is not likely to occur under most soil and plant conditions.

Micronutrients

The relative response of the tomato plant to the micronutrients has been given by Lorenz and Maynard (1988) and Vitosh et al. (1994). In Florida on virgin sandy soils, or on sandy soils where a proven micronutrient need exists, the rates of application recommended (Hochmuth, 1997) are

Micronutrient	Pounds per Acre (lb A^{-1})
Boron (B)	2.0
Copper (Cu)	2.0
Iron (Fe)	5.0
Manganese (Mn)	3.0
Molybdenum (Mo)	0.02
Zinc (Zn)	2.0

In Georgia, a general application of 1 lb B A^{-1} is recommended; and for soils low in Zn, 5 lb Zn A^{-1} is recommended.

Although the micronutrients are essential elements, their deficiency is not very common in most soil situations. If a micronutrient insufficiency is suspected, it should be based on determinations obtained by means of a soil test or plant analysis, or both. Since most of the micronutrients can be toxic to the tomato plant when in excess, an indiscriminate micronutrient application is not recommended, whether as a specific application or as an application by means of a micronutrient-supplemented fertilizer. For more details on the micronutrients and their use, the books edited by Mortvedt (1991) and written by Pais and Jones (1997) are excellent resources.

NUTRIENT CONTENT AND REMOVAL

Harvested tomato fruit and vines can remove sizable qualities of N, P, and K from the soil. Lorenz and Maynard (1988) reported that a 600 cwt A^{-1} tomato crop will contain:

Plant Part	N (lb A^{-1})	P (lb A^{-1})	K (lb A^{-1})
Fruit	100	10	100
Vines	80	11	100

Halliday and Trenkel (1992) reported that a 24 ton ha^{-1} tomato crop will remove 117, 46, 319, 43, and 129 kg ha^{-1} of N, P_2O_5, K_2O, MgO, and Ca, respectively.

One means of establishing the basis for a fertilizer recommendation is to determine the elemental content of the crop and then apply a similar rate of that element as fertilizer to meet that elemental crop demand. Such a determination is given in Table 5.5 for both an outdoor and a greenhouse tomato crop.

TABLE 5.5
Nutrient Demand, Uptake, and Removal

	Kilograms per Hectare (kg ha^{-1})			
	N	P_2O_5	K_2O	MgO
Outdoor crop (40–50 ton ha^{-1} yield)	100–150	20–40	150–300	20–30
Greenhouse (100 ton ha^{-1} yield)	200–600	100–200	600–1000	—

Source: Halliday, D.J. and M.E. Trenkel (Eds.). 1992. *IFA World Fertilizer Use Manual*, pp. 289–290, 331–337. International Fertilizer Industry Association, Paris, France.

Note that in this table, the level of elements given are related to a specific crop production level. If the yield is less or greater, the level of element demand, uptake, and removal would be somewhat different.

DRIP IRRIGATION AND FERTIGATION

Control of the water and nutrient element supply to the plant is becoming increasingly important to obtain high yields and high fruit quality. Geisenberg and Stewart (1986) have given the amount of water needed in field tomatoes if no rain falls as 2,000–6,000 m^3 ha^{-1} and under extremely dry conditions as 8,000–10,000 m^3 ha^{-1}. If the water supplied to the plant can be controlled, the yield and the total dissolved solids (TDS) of the fruit can be also controlled. Such control is accomplished by using some form of plasticulture, a system of production that was discussed in a special issue of the journal *HortTechnology*, volume 6, issue 3, pages 150–192, 1996. Geraldson (1963, 1982) was able to precisely manage water and nutrient element supply with his system of growing on soils where the height of the underlying water table could be controlled.

For growing tomatoes in a plasticulture system, a raised soil bed is made, which is covered with plastic; and then a drip irrigation system is placed down the row inside the plastic cover (Clark and Smajstrla, 1986a; Hartz, 1986). A small hole is cut into the plastic and a tomato transplant is put into the bed soil alongside the drip line (Orzolek, 1986). The components for such a system have been described by

Lamont (1996). The plastic cover keeps rainfall off the soil and the grower is able to maintain the water and nutrient elements at their optimum levels by means of the drip irrigation system into which fertilizer has been added to the irrigation water (Clark and Smajstrla, 1996b).

Resh (1995) has described the various drip irrigation systems that are used in greenhouse production systems.

The use of trickle (drip) irrigation for controlling water and the delivery of fertilizer to a crop has been discussed in detail in the books by Keller and Bliesner (1990) and Nakayama and Bucks (1986), and in the extension bulletin by Snyder (1996c).

SOIL TESTING

A soil test is the only means for determining the chemical characteristics of a soil, and growers should be routinely testing to determine what that status is and what effect soil amendments and cropping are having on the chemical characteristics of the soil. A routine soil test will normally include pH; and level of extractable P, K, Ca, and Mg. Some laboratories may also include the micronutrients, B, Fe, Mn, and Zn, as a part of the routine test, or only determined by request. Other soil parameters, such as texture, soluble salt level, and organic matter content, may or may not be included in the routine test or even in the requested category.

For some, only the test values will be requested while others may expect or will request an interpretation of the test results plus a recommendation. The two books, one edited by Westerman (1990) and the other by Carter (1993), describe the soil testing technique that includes sampling and laboratory testing procedures, The Westerman (1990) book also includes chapters on soil test data interpretation and application.

The time (usually done several months prior to soil preparation, or taken when plant tissue is collected for a plant analysis), method of sampling (i.e., soil depth—surface or plow), coring pattern (random), and number of cores (from 20 to 40 per unit area) needed to obtain a representative sample are given in some detail by Carter (1993). In addition, most soil testing laboratories will provide information on how to obtain a representative sample for testing.

It is always wise to make contact with the laboratory being asked to perform the test before collecting and submitting samples. The laboratory selected should be familiar with the environmental or soil characteristics of the area from which soil samples are to be taken. A list of soil testing laboratories can be found in two works, one by Downing and Associates (Downing, 1997) and the other published by CRC Press (Anon., 1998b).

PLANT POPULATION AND SPACING

COMMERCIAL PRODUCTION

Plant densities range from 12,150 to 36,900 plants per hectare depending on plant type, whether determinate or indeterminate, the former type mainly for processing and the latter for fresh market. The suggested spacings between plants within the row and between the rows are

	In Row (in.)	Between Row (in.)
Staked	12–24	36–48
Processing	2–10	42–60

In a plasticulture system using drip irrigation and fertigation, the recommended in-row plant spacing in a single row system was 18–24 in. (Lamont, 1996).

Halliday and Krenkel (1992) report a plant density of three to four plants per square meter or a field size density of 12,150–36,900 plants per hectare. Rubatzky and Yamaguchi (1997) report that processing tomatoes grown on beds, 150–180 cm wide with in-row spacing from 30 to 60 cm, result in a population of 10,000–20,000 plants per hectare. For fresh market fruit, plant populations range from 8,000 to 14,000 plants per hectare; if staked, the population would be 6,000–8,000 plants per hectare. Typical spacing would be 60–75 cm within the row and 120–150 cm between rows.

For staked tomatoes, Peet (1996d) suggests 15–30 in. between plants in the row and 54–72 in. as the between-row spacing. Spacing selections will depend on the vigor of the cultivar grown and whether the plants are or are not pruned. For processing, two to four plants may be clumped together with 7–12 in. between plants in 5-ft double or single rows, with no more than five plants per foot of row.

Much of the variability in row spacing is frequently due to the type of equipment available that is used to prepare the soil, set the transplants, and cultivate.

Home Garden Spacing

For the home gardener, Burnham et al. (1996) give a range of 35–65 plants per 100 ft of row with 18–36 in. between plants, the wide range in spacing depending on how the plants are managed. Staked plants would require less space with as many as 65 plants per 100 ft of row with 18 in. between plants. Wider spacing between plants in the row or between the row provides accommodation for easy movement among the plants. For unstaked plants, a wider space within the row and between rows would be required to keep plants from making contact with each other as well as allowing for movement within the garden for controlling weeds and harvesting fruit.

Some growers put the plant within a round wire cage, allowing suckers to grow and produce fruit. In such a system, sufficient space must be provided for ease of harvest and maintaining the area around the plants.

FRUIT YIELD PRODUCTION

Commercial

Yearly fruit yields reported for the years from 1985 to 1996 (USDA, 1997) ranged from a low of 250 cwt A^{-1} in 1987 to a high of 296 cwt A^{-1} in 1992 for fresh market, and a low of 29.19 ton A^{-1} in 1990 to a high of 33.94 ton A^{-1} in 1994 for processing tomatoes. The wide range in yields reflects the seasonal variability and climatic factors that impact plant growth and fruit yield. Maynard and Hochmuth (1997) give the following as average and good fruit yields for three different types of tomatoes:

Type	Average (cwt A^{-1})	Good (cwt A^{-1})
Fresh market	280	410
Processing	650	900
Cherry	—	600

In a plasticulture system using drip irrigation and fertigation, Lamont (1996) reported seasonal fruit yields of 3200 boxes (20 lb per box) under Florida conditions.

Halliday and Trenkel (1992) reported potential yields for fresh market tomatoes as 40–50 ton ha^{-1} and for processing tomatoes, 20–60 ton ha^{-1}.

Home Gardening

Fruit yield will depend on the plant type (determinate or indeterminate), maturity of the variety, environmental conditions, and skill of the gardener. On the average, per plant fruit yields would be expected to range between 4 and 10 lb per plant. Burnham et al. (1996) gave a 100 ft of row (35–65 plants) yield of 125 lb. Cavagnaro (1996) has described a home garden procedure for producing tomatoes by the ton. The video, *Tons of Tomato*, is available from *The New Garden Journal* (Walls, 1980).

HOME GARDEN PRODUCTION PROCEDURES

Tomato is the most commonly grown vegetable in home gardens in the United States. Raymond and Raymond (1978) have described the various procedures that are required to grow tomatoes in home garden situations. In the 1976 Ortho Book—Southern Edition (Ray, 1976), *All About Tomatoes*, the following needs were listed:

- A continuous supply of moisture
- A continuous supply of nutrients
- Air in the soil for healthy growth
- About 8 h of daily sunlight
- Night temperatures (at least during part of the night) that permit flowering and setting of fruit
- Protection from insects, diseases, and dogs that dig holes to nap in

Although the tomato plant can grow well in a range of soil textures, sandy soils or those soils that tend to be droughty are the least desirable unless the soil moisture level can be maintained to avoid plant stress. The occurrence of BER is more likely on such soils than on soils that have a substantial water-holding capacity. Since the tomato plant cannot tolerate waterlogged soils, those that are poorly drained should be avoided. For best results, the garden soil should be irrigated to maintain as constant a moisture level in the soil as possible.

Soils that are relatively high in organic matter content are desirable because humus, the end product of organic matter decay, can contribute to the soil's water-holding capacity; and in addition the decay of organic material in the soil will release substantial quantities of several important elements, mainly N, P, and B, into the soil solution for utilization by the tomato plant. Parnes (1990) describes what constitutes a fertile soil, giving a guide to organic and inorganic fertilizers.

The tomato plant grows best in soils with a pH ranging between 6.0 and 6.5. If the soil water pH is below 6.0, then agricultural limestone—dolomite—is the preferable liming material because it contains Mg. This limestone should be added and mixed to the plow or spading depth of the soil several months prior to fertilizing or transplanting. An acid (pH 5.5–6.0) soil will require from 10 to 25 lb of limestone per 1000 ft^2, with one application sufficient to maintain the soil pH for 2 to 3 years.

In general, soils testing at the "moderate" to "high" levels are best for the tomato plant. A typical fertilizer recommendation for the home garden tomato at varying soil test levels of P and K are as follows:

Soil Test Level	Fertilizer Treatments
"Low" for P and K	Broadcast 45 lb of a 6–12–12 fertilizer per 1000 ft^2, or apply 16 lb of a 6–12–12 fertilizer per 100 ft of row
"Medium" for P and K	Broadcast 36 lb of a 6–12–12 fertilizer and 3 lb of ammonium nitrate per 1000 ft^2, or apply 12 lb of a 6–12–12 fertilizer plus 1 lb of ammonium nitrate per 100 ft^2 of row
"High" for P and K	Broadcast 28 lb of a 10–10–10 fertilizer or 35 lb of a 8–8–8 fertilizer per 1000 ft^2, or apply 10 lb of a 10–10–10 fertilizer or 12 lb of a 8–8–8 fertilizer per 100 ft^2 of row

These fertilizer grades and application rates are given just as examples of what would be required based on three soil test levels for P and K. Home gardeners are advised to have their garden soil tested in the fall or very early spring by a soil testing laboratory in their area, and then to carefully follow the recommendation. The fertilizer should be applied prior to transplanting and not at the time of transplanting so that the fertilizer has time to interact with the soil. Errors made in applying too much or too little fertilizer can have a significant effect on tomato plant growth and fruit production; therefore, care should be exercised when applying any fertilizer. Plants that are nutritionally sound are less likely to be affected by short periods of adverse climatic conditions as well as being more resistant to pests.

Many home gardeners apply compost and other types of "natural" organic materials to their garden soils to improve soil tilth and to add naturally occurring fertilizer elements. Care should be used to ensure that what is added is free from pest chemicals (tomato is particularly sensitive to herbicides) and not imbalanced in terms of elemental content. Procedures for managing a garden soil, organically as well as inorganically, are given in the book by Parnes (1990).

The production and handling of tomato plant seedlings (transplants) are covered in Chapter 7 of this book.

FIGURE 5.1 Placement of tomato plants around a cage of organic compost. (From Tonge, P. 1979. *The Good Green Garden.* Harpswell Press, Brunswick, ME. With permission.)

ORGANICALLY GROWN

There is considerable interest in the organic production of vegetable crops, following practices that do not include the use of chemically derived fertilizers and pest chemicals (Bartholomew, 1981). For the home gardener, the use of natural products to sustain the fertility level of the soil and to control insect and disease pests is possible from a practical standpoint (Harris et al., 1996; Jesiolowski, 1996), although more demanding in terms of soil and plant management (Parnes, 1990). Instructions for growing plants by "organically based procedures" are given in the books by Hendrickson (1977) and Tonge (1993), and are found in issues of the *Organic Gardening* magazine (Rodale Press, Emmaus, PA).

One system of growing is to take fencing wire 3–4 ft high and make a circle 3–5 ft in diameter, setting plants around the wire circle at 2-ft intervals about 6–12 inches from the circle. Fill the wire cage with well-rotted compost and manure or other organic materials to a depth of 6 in. (Figure 5.1). Water is applied to the compost in the cage, and the flow of nutrients from the decomposing material in the cage will supply the plants around the cage circle with the nutrients needed. This system of growing is described in more detail by Tonge (1979). It should be remembered that the basic fertility of the soil and the source and composition of the organic materials added to a soil will have a significant effect on the results obtained with this system.

For large-scale production, growing plants organically can be difficult. What can be termed *organic* in terms of the product produced is not well defined because states vary considerably in their requirements for such a designation. The establishment of a federal standard for organically grown products is being considered by the United States Department of Agriculture (USDA), and news articles on this effort have been published by *The New Garden* (Anon., 1998d), *Organic Gardening* (Anon., 1998e), and *Ag Consultant* (Melnick, 1998). The National Organic Program (NOP) calls for:

- Establishment of production and handling standards for organic foods
- Program to accredit state and private agencies wishing to certify farms or processors under the organic program
- Labeling requirements for all organic products
- Enforcement provisions
- Approval program for imports with equivalent requirements

Once the NOP is finalized, organic growers and processors will earn a USDA seal under the guidelines. To be certified, farms must have a minimum 3-year history free of prohibited substances. Raw products must be 100% organic, while processed foods must be 95% organic, which allows for a small amount of ingredients not available organically. Processed foods between 50 and 95% organic may be labeled as containing "certain organic ingredients."

For those interested in following the issues associated with the use of organic procedures for crop production can obtain the latest information on the worldwide web: <www.iquet.net/ofma> the web site of the Organic Farmers Marketing Association. The web site for Rodale Press, who publishes *Organic Gardening,* is <http://www.rodalepress.com/organicrules.htm>.

6 Greenhouse Tomato Production

CONTENTS

CURRENT AND FUTURE STATUS

The status of the greenhouse tomato industry in northwestern Europe up to late 1985 has been reviewed by van de Vooren (1986), with slightly over 6,500 ha of greenhouses producing some 940,000 tons of fruit each year. Wittwer and Castilla (1995) reported that 279,000 ha (689,409 A) of greenhouses were being devoted to vegetable production worldwide. In 1992, the value of Dutch-produced greenhouse vegetables was US$1.6 billion (Ammerlaan, 1994); in Canada, US$98 million (Statistics Canada, 1993); and in the United States, US$31.7 million (Snyder, 1993a). Although these figures represent all vegetables, tomato, cucumber, lettuce, and pepper, that are commonly grown in greenhouses, the majority of the product being produced is tomato. All these figures show the enormous growth that has continued for the greenhouse industry. Curry (1997) has reported on the $35 million greenhouse tomato industry in Colorado, which produces 32 million lb of fruit annually on 68 covered acres at three locations.

Good review sources are the proceedings edited by Savage (1985) and that published by ASHS Press (Anon., 1996).

Initially a number of organic-based substrates were used, even soil; however with the introduction of rockwool, this substance has become the primary growing media in widest use around the world for the production of greenhouse-grown tomatoes (Johnson et al., 1985; van de Vooren et al., 1986; Logendra and Janes, 1997).

The acreage of greenhouses devoted to vegetable production by various countries is given in Table 6.1.

TABLE 6.1
Estimates of Greenhouse Tomato Acreage in Various Countries

Country	Acres (Hectares)
Canada	710 (287)
England/Wales	3,000 (1214)
The Netherlands	11,400 (4613)
Spain	30,000 (12,140)
United States	450 (182)

Source: Snyder, R.G. 1996a. In: *Greenhouse Tomato Seminar*. ASHS Press, American Society for Horticultural Science, Alexandria, VA.

The continuing expansion of greenhouse tomato production will be influenced by many factors, economic, political, and environmental as well as a sustained consumer demand for fruit year round. The ability of the current transportation system to move large quantities of fruit economically from one continent to another has brought greenhouse-grown fruit to almost every corner of the world. The impact of genetics and genetic engineering, which can develop cultivars with higher yield potentials and fruit quality characteristics than what exists among current cultivars in use today, is probably the most important factor that will determine how large the greenhouse industry will become in the future.

Increased production per unit of greenhouse space by the application of improved technology through cultivar development and control of the greenhouse environment has been considerable in the past two decades (Snyder, 1996a). For example, per plant fruit yield and yield per unit of space and time have almost doubled, and that increase in production efficiency is expected to continue. Morgan (1997) reported that fruit yields of greater than 50 kg m^{-2} (88.2 lb yd^{-2}) have been obtained representing the higher production capability, while mediocre fruit yields were 10–20 kg m^{-2} (17.6–35.3 lb yd^{-2}) per year. Halliday and Trenkel (1992) gave 100 ton ha^{-1} as the fruit yield for greenhouse production, which compares with their figure of 20–60 ton ha^{-1} for field-grown fresh market tomatoes. Mizra (1994) has discussed the requirements for managing greenhouse production to obtain above average yields.

It is expected that the acreage of greenhouses devoted to tomato production will continue to increase, particularly in those regions of the world where consumer demand is high and in areas where environmental conditions would favor sheltered production methods (Wittwer, 1993; Janes, 1994; Jensen and Malter, 1995).

Although the acreage devoted to greenhouse tomato production in the United States has lagged behind that in many other countries, there is considerable interest in this method of production. For a comparison, the HSA Proceedings edited by Wignarajah (1997) should be compared to that edited by Savage (1985).

The leading states in the United States in greenhouse tomato production from two sources are listed in Table 6.2.

TABLE 6.2
Leading Greenhouse Tomato Production States in the United States

State	Acres[a]	Acres[b]
Arizona	25	44
California	8	30
Colorado	69	94
Florida	10	7
Mississippi	15	16
New Jersey	—	15
Nevada	—	12
New York	35	35
North Carolina	10	10
Ohio	15	20
Pennsylvania	49	56
Tennessee	—	20
Texas	—	72

[a] Snyder, 1996a.
[b] Naegely, 1997.

Brentlinger (1997) also has made a determination of the acreage of greenhouse tomato in the United States, looking at the current status and future expansion for 1997 as is shown in Table 6.3.

TABLE 6.3
Summary by State of the Large Tomato Growers in the United States and Their Current Acreage, along with Projected Acreage to Be Installed in 1997

State	Current Acreage	New Acreage in 1997	Total
Arkansas	45	25	70
California	20	—	20
Colorado	70	80	150
Nevada	10	—	10
New York	10	—	10
Pennsylvania	30	—	30
Texas	40	80	120
Total	225	185	410

Source: Brentlinger, D. 1997. pp. 67–73. In: R. Wignarajah (Ed.), *Proceedings 18th Annual Conference on Hydroponics*, Hydroponics Society of America, San Ramon, CA.

The viability and expansion for greenhouse tomato production in the developed world, such as the United States, is based on the ability of growers to:

- Maintain a constant supply of high quality fruit to the marketplace at competitive prices
- Provide pesticide free product
- Maximize production capacity through precise space utilization while minimizing costs of production

The growth of the greenhouse industry in the United States in just the past few years suggests that all these preceding factors are attainable.

FACTORS AFFECTING GREENHOUSE PRODUCTION

The major factors that affect greenhouse tomato production are

- Light, both intensity and length
- Carbon dioxide (CO_2) level in the greenhouse
- Temperature and humidity control, both low and high
- Disease and insect control
- Nutritional management over the life of the tomato plant
- Varietal plant characteristics
- Management skill required to produce higher plant yields of quality fruit

An excellent review of some of these factors has been written by Papadopoulos et al. (1997), factors that are discussed in detail in Chapter 2. Tite (1983) authored a simple guide for greenhouse tomato production covering all aspects of production, basic procedures that are still applicable today. Hochmuth (1991) describes greenhouse tomato production under Florida conditions; Snyder (1997a), for the state of Mississippi; and Curry (1997), for the state of Colorado.

In the past 10 years, there have been a number of very significant developments that have affected the ability to produce high quality fruit in a greenhouse-controlled environment, which include:

- There has been a change from growing in soil to some form of soilless production, such as Nutrient Flow Technique (NFT) hydroponics, or perlite bag or rockwool slab–drip irrigation systems (Resh, 1995), which provides a degree of nutrient element control not possible in soil and eliminates soil factors that are difficult to control (Wittwer and Honma, 1969).
- Cultivars have been bred specifically for greenhouse conditions and low light situations, having either resistance or tolerance to common tomato plant diseases and insects; and having significantly increased fruit yield potentials, producing fruit with specific fruit characteristics to meet particular consumer preferences (Waterman, 1993–1994; Baisden, 1994; Morgan, 1997; Pierson, 1997)
- Introduction of bumblebees for flower blossom pollination eliminates the need to hand pollinate, a major labor-intensive operation (Gunstone, 1994; Kueneman, 1996).
- Use of predator insects (Hussey and Scopes, 1985; Malais and Ravensberg, 1992) and other nonchemical techniques can either eliminate or significantly reduce the need for chemicals to control plant-damaging insects and disease as well as integrated pest management (IPM) procedures (Shipp et al., 1991; Clarke et al., 1994; Ferguson, 1996; Kueneman, 1996; Waterman, 1996; Peet, 1996a, 1996b; Johnson, 1997; Papadopoulos et al., 1997).
- Computer control of the growing system and greenhouse environment is based on factors being continuously and automatically monitored (Gieling, 1985; Bauerle et al., 1988; Bauerle, 1990; McAvoy et al., 1989a, 1989b; Giacomelli and Ting, 1994; Edwards, 1994; Giacomelli, 1996a, 1996b; Snyder, 1996b; Giniger et al., 1998; Hanan, 1998).

A decision model for hydroponic tomato production (HYTODMOD) was developed for achieving high yield quality fruit by Short et al. (1997) who identified five key tests as the hydrogen ion concentration and the electrical conductivity of the feeding solution, root temperature, greenhouse air temperature, and relative humidity. The five growth stages were

- Germination and early growth until roots emerge
- Seedling growth until transplanted into growing media

- Vegetative growth until first flower opens
- Early fruiting starting from the time first flower opens until first fruit is picked
- Mature fruiting starting from time first fruit is picked until crop is terminated

Ranges of risk for the five tests were developed and the HYTODMOD program was validated by comparing model recommendations with four experts who were given 25 hypothetical and random production situations, the following being the optimum ranges for the following factors:

Factor	Growth Range	Optimum Range (°C)
Air temperature	Germination to seedling stage	24–26
Sunny daytime	Seedling to termination	24–27
Cloudy daytime	Seedling to termination	22–24
Night air temperature	Seedling to termination	18–20
Root temperature	Germination to early growth	24–27
	Vegetative to termination	20–24
		(%)
Relative humidity	Germination to early growth	75–88
	Seedling stage	70–80
	Vegetative to termination	60–80
pH nutrient solution	Germination to early growth	5.5–6.5
		($dS\ m^{-1}$)
Electrical conductivity of nutrient solution	Germination to early growth	1.8–2.0
Sunny day	Seedling to termination	1.5–2.0
Cloudy day	Seedling to termination	2.5–4.0

Similar studies related to greenhouse production need to be made to evaluate all those aspects of production, ensuring that current recommendations are suitable to maintain and increase the productive capability, and to determine those parameters than need to be further investigated and improved.

GREENHOUSE STRUCTURES

The greenhouse grower has a variety of greenhouse structures to choose from in terms of size and covers, the use of plastic-covered greenhouses being increasingly used worldwide (Goldberry, 1985; Wittwer, 1993; Jensen and Malter, 1995; Snyder, 1996b, Giacomelli 1996a; Hanan, 1998). For the single owner–operator, a standard-sized greenhouse would be 30–40 ft wide and 100–140 ft in length, with the cover being a single or double air-separated layer of plastic film (Figure 6.1). For larger installations, greenhouses are normally gutter-connected with combinations of covers being either totally plastic or glass, or a mix of glass and plastic film or sheets of fiberglass or plastic (Figure 6.2). Giacomelli and Ting (1994) have written a

FIGURE 6.1 Polyethylene-covered greenhouse with white shade cloth in place.

FIGURE 6.2 Glass-covered greenhouse suitable for tomato production.

bulletin on design characteristics of greenhouses suitable for the production of tomatoes, and the book by Hanan (1998) provides details on the construction and operation of a greenhouse. Details on greenhouse structures and systems have been reviewed by Goldberry (1985).

The size and design of the installed heating and cooling systems will vary considerably depending on location (latitude). In northern latitudes, an efficient heating system is the dominate requirement (Papadakis et al., 1994), while in southern latitudes, cooling efficiency is the dominant requirement (Hochmuth and Hochmuth, 1996). The heating and cooling system must be able to maintain an optimum air temperature within the range of 70–75°F (21–24°C), keeping the minimum temperature from dropping below 65°F (18.3°C) and the maximum temperature from exceeding 85°F (29.4°C) (Giacomelli, 1996b). Snyder (1996b) has described what controls are needed to properly maintain the temperature within a greenhouse. Some provision for shading the greenhouse during periods of high light intensity is also required, even in the northern latitudes if fruit production is to continue through the summer months. Floor heating is proving to be very advantageous in keeping the rooting media from dropping below the optimum rooting temperature of 70°F (21°C).

Air movement within the greenhouse is important, with warm dry air introduced at the bottom of the plant canopy so that air flow is from the base of the plant up through the canopy into the greenhouse gable. The objective is to keep the plant canopy as dry as possible, which prevents the development of diseases and a potential habitat for insects in the older, lower foliage. Also such an air flow system prevents layering from occurring as photosynthesis lowers the carbon dioxide (CO_2) content of the air trapped within the plant canopy (Harper et al., 1979). Without mixing, the plant's photosynthetic rate within the plant canopy would decline. Maintenance of the CO_2 content of air in the greenhouse at its normal ~330 ppm level is achieved by air mixing as well as replenishing air in the greenhouse with outside air, or by adding CO_2 to the air in the greenhouse (Knecht and O'Leary, 1974; Harper et al., 1979; Willits and Peet, 1989; Tripp et al., 1991). Schwarz (1997) has warned of the potential for CO_2 toxicity, giving toxic levels and plant symptoms. The effect of CO_2 supplementation is discussed in more detail in Chapter 2.

In addition, free water should not be present in the greenhouse with all open surfaces being kept as dry as possible. The ideal relative humidity is 50%, with the acceptable range being from 40 to 70%.

SITE LOCATION

The location of the greenhouse, its orientation (north and south longitudinal orientation recommended), and surrounding environment are important considerations because site selection can determine the difference between success and failure. Before selecting the greenhouse site, long-term weather records should be examined to determine hours of sunshine, numbers of days with varying cloud cover, wind direction and speed, and extremes of air temperature.

Placement of the greenhouse downwind of a major agricultural area or sizable industry that may be introducing gases or dust into the atmosphere should be avoided. The tomato plant is quite sensitive to a number of atmospheric substances that are listed in Table 6.4.

TABLE 6.4
Sensitivity of Tomato Plants to Air Pollutants

Pollutant	Sensitivity
Ozone (O_3)	Sensitive
PAN	Sensitive
Ethylene (C_2H_4)	Sensitive
2,4-D (herbicide)	Sensitive
Hydrogen sulfide (H_2S)	Sensitive
Sulfur dioxide (SO_2)	Intermediate
Chlorine (Cl_2)	Intermediate
Ammonia (NH_3)	Intermediate
Mercury (Hg) vapor	Intermediate

Source: Lorenz, O.A. and D.M. Maynard. 1988. *Knott's Handbook for Vegetable Growers*. 3rd ed., John Wiley & Sons, New York.

A suitable windbreak can reduce heating costs and provide some degree of protection from foreign substances making contact with the greenhouse surface or from being drawn into the greenhouse through its ventilation system.

The immediate area around the greenhouse should be kept clean and free of untended plants. If grass is grown around or near the greenhouse, it should be kept free of weeds and frequently mowed. A dust-free environment around the greenhouse should be maintained at all times.

WATER QUALITY

A considerable amount of high quality water is needed, the volume determined by the number of plants being grown and method of growing. Water quality is determined by its source and freedom from suspended and dissolved substances. Pure water, although the most desirable, is usually not readily available. An abundant supply of domestic drinking water is not a guarantee as a quality source since it too may contain substances that can affect plants adversely. Therefore, no matter what the water source, it should be tested to determine its content of elements and organic substances; and if necessary, a procedure should be worked out for removing undesirable substances (Anon., 1997a).

Elements or ions found in many water sources that are undesirable are boron (B), sodium (Na), chloride (Cl^-), chlorine (Cl_2), sulfide (S^-), fluoride (F^-), carbonate (CO_3^{2-}), and bicarbonate (HCO_3^-). If these elements and ions are present at substantial concentrations, they must be removed or diluted to such a level that they will not affect plants. A guide as to what maximum level these elements or ions can be in irrigation water to be used in a rockwool–drip irrigation system, for example, has been given by Verwer and Wellman (1980):

Element/Ion	Maximum Concentration ($mg\ L^{-1}$, ppm)
Chloride (Cl)	50–100
Sodium (Na)	30–50
Carbonate (CO_3)	4.0
Boron (B)	0.7
Iron (Fe)	1.0
Manganese (Mn)	1.0
Zinc (Zn)	1.0

Farmhand et al. (1985) have also given guidelines for irrigation water quality characteristics in terms of the *degree of problem* as:

Characteristics	Degree of Problem		
	None	Increasing	Severe
EC, dS m^{-1}	<0.75	0.75–3.0	>3.0
TDS, mg L^{-1}	<480	480–1920	>1920
Sodium (Na), SAR value	<3	3–9	>9
Chloride (Cl), mg L^{-1}	<70	70–345	>345
Boron (B), mg L^{-1}	1.0	1.0–2.0	2.0–10.0
Ammonium (NH_4) and nitrate (NO_3), mg L^{-1}	<5	5–30	>30
Bicarbonate (HCO_3), mg L^{-1}	<40	40–520	>520

Note: EC, electrical conductivity; TDS, total dissolved solids.

Calcium (Ca) and magnesium (Mg), two essential elements for plants, are commonly found in many water supplies; and when present, they can provide a portion or most of that needed to sustain plant growth. For those using a complete nutrient solution to supply plants with the essential elements, these elements may be in sufficient concentration in the water to reduce or eliminate the need to add reagents for supplying these elements, resulting in substantial savings in chemical costs. However, failure to compensate for their presence can lead to elemental imbalances and stressed plants.

With the continuous use of water containing a substantial concentration [>30 mg L^{-1} (ppm)] of Ca (frequently referred to as *hard water*), the pH of the growing media will increase and eventually become alkaline. For example, over many years of use of Ca-containing well water [50–80 mg L^{-1} (ppm)] in the soil medium tomato greenhouses in the Cleveland, OH area, the soils became alkaline (pH 8.3); they contained excess calcium carbonate ($CaCO_3$) and calcium sulfate ($CaSO_4$) at soil levels, which posed a serious problem to the grower. Growers were forced to find another source of water, while dealing with the adverse effects due to the created soil alkalinity.

Organic substances, such as pest chemicals and petrochemicals, and suspended organic substances, must be removed before the selected water supply would be suitable for greenhouse use.

Water treatment to remove undesirable substances can be expensive, requiring filtering to remove suspended substances, carbon-type filtering to remove organic chemicals, and ion exchange or reverse osmosis treatment to remove inorganic ions (Anon., 1997a). The treatment system must be of sufficient size and capacity to provide the volume of clean pure water needed within the time frame required.

The release of spent water or nutrient solution from the greenhouse may require control measures depending on local environmental laws and regulations because these effluents will contain nitrate (NO_3) and phosphate (PO_4) ions, ions that can contribute to surface or groundwater pollution.

CULTURAL PRACTICES

PLANT SPACING AND DENSITY

According to Papadopoulos (1991), the optimum space per plant is 0.35–0.40 m^2, planted in double rows at 80-cm spacings with 1.2 m between the double rows. Snyder (1997a) suggests 4 ft^2 per plant for a population of 10,000 plants per acre. The arrangement is double rows about 4 ft apart with 14–16 in. between plants in the row.

With the plant physical arrangement being normally in double rows, the space between the rows and the plant spacing within the row can significantly affect light penetration through the canopy. Harper et al. (1979) found that with a plant spacing of 2.5 plants per square meter in a plant spacing configuration of 45 × 45 cm, about 30–70% of the solar radiation during the 1-h period of solar noon reached the greenhouse floor. By increasing the plant spacing to 3.5 plants per square meter, 60–70% of the solar flux was intercepted by the plant canopy and fruit yield was affected for a double row of plants spaced 18 in. (45 cm) apart. In a similar evaluation, Cocksull et al. (1992) studied the influence of shading on yield, shading that can be varied by altering the plant spacing pattern within the greenhouse. Papadopoulos and Pararajasingham (1996) evaluated plant spacing on the basis of its impact on photosynthesis. There is no *ideal* plant spacing pattern best suited for every growing system. The object is to utilize all the growing space to maximize light interception and provide sufficient space between rows to service the plants.

PLANTING SCHEDULES

A two-crop season is the common practice in the northern latitudes: planting is done in late August and harvesting through December; and then planting in March and harvesting into June or July, avoiding the cold low light intensity periods. In the lower latitudes where the period of cold and low light intensity is less, a single crop planted in September and carried through until June or July is commonly practiced. Curry (1997) describes the system used in Colorado greenhouses, a staggered schedule of transplanting and replacing plants every 8 months in a two-row planting system that ensures continuous production of fruit year-round.

FIGURE 6.3 A sucker will appear at the leaf axil (between the main stem and leaf petiole) and if not removed will produce another flowering stem.

Pruning and Training

Maintaining a plant through a long growth period can be a challenge to the grower. The plant is trained up a single vertical plastic twine, with suckers removed (Figure 6.3) to maintain a single stem (Figure 6.4). When the top of the plant reaches the horizontal support wire to which the vertical support twine is attached, the plant can be either topped (terminating further extension of the plant) or lowered (allowing plant growth to extend up the lowered twine).

Sanitation Practices

Sanitation in the greenhouse is essential. Leaf material, such as suckers, immature–misshapen fruit, and senescent leaves, must be immediately removed from the greenhouse to keep the greenhouse floor and growing area completely free of discarded plant materials or other foreign substances. Plant material that is infested with insects or diseased must be immediately removed from the plant and taken from the greenhouse and destroyed. Any tools or devices that will come in contact with plants should be carefully cleaned and sterilized after use or before reuse.

Tomato Plant Maintenance

Careless management of the plants in the greenhouse can be costly in terms of lost yield and low fruit quality. Daily attention is required to ensure that plants are kept upright on the suspension growing system, suckers are promptly removed when appearing, abnormal looking plant material is promptly removed, fruit clusters are

FIGURE 6.4 Single-stem tomato plant on support line and showing cluster support devices attached.

pruned to the desired number, and abnormal fruit is removed when first appearing. Any abnormal growth or plant appearance should be carefully evaluated, questionable-looking plants should be removed from the greenhouse, and steps should be taken to determine the cause for any plant abnormality that appears.

Flower Pollination

The formation of pollen and its transfer to the stigma are discussed in detail by Ho and Hewitt (1986). Although all current varieties are self-pollinated, the transfer of pollen to the stigma under greenhouse conditions may not occur in such a manner than would ensure complete pollination, thereby resulting in poorly shapen fruit.

If the flower blossoms are hand pollinated, flower vibration using a mechanical vibrator (see Figure 2.1) must be done daily based on a preplanned program following the correct procedure to keep from damaging emerging fruit.

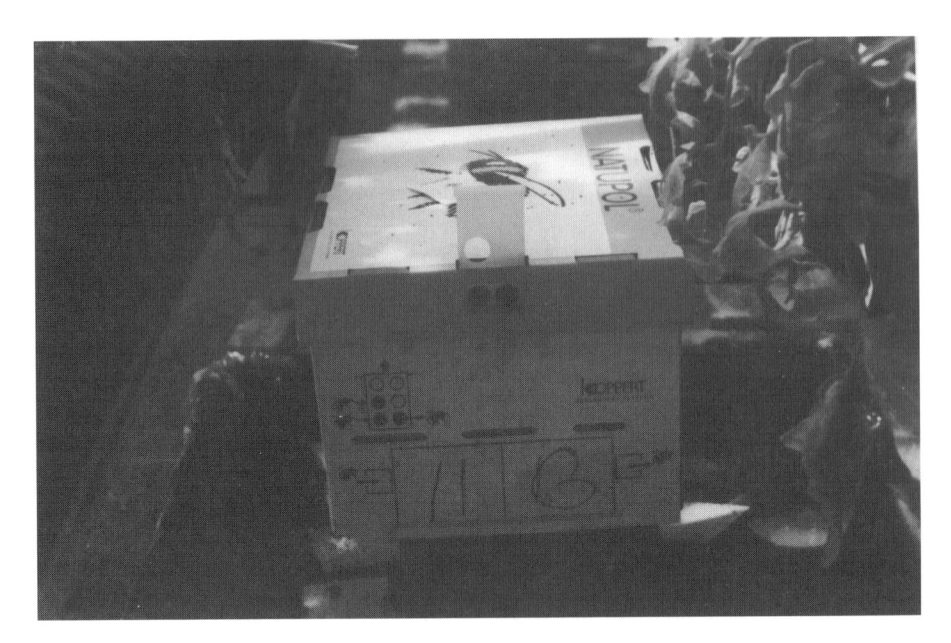

FIGURE 6.5 Bumblebee hive placed in greenhouse to provide bees for tomato flower pollination.

If bumblebees are being used for pollination, the time of placement and size of the hive (Figure 6.5) are determined by the stage of plant development and number of plants (Gunstone, 1994). If few flowers are ready for pollination, repeated pollination visits by bees to a flower may result in damage to that flower. Flowers that are not pollinated will fall from the cluster stem whereas incomplete pollination will result in misshapen fruit.

Cluster Pruning

Cluster pruning, practiced to maintain fruit size, is the removal of flowers or small fruit to keep a set number of fruit per cluster (Cocksull and Ho, 1995). However, severe fruit pruning can result in blossom-end rot (BER) on some fruit (DeKock, 1982). Any fruit that is not normal in shape or has been physically damaged should be removed from the plant when once observed. Fruit removed from the lower trusses results in an increase in fruit size on the upper trusses.

WORKER SKILLS AND HABITS

Skilled greenhouse workers are required, workers that have been well trained and knowledgeable as to the tasks required to maintain productive plants. Control of worker access into the greenhouse is important to maintain the greenhouse free of disease and insect pests. Depending on local conditions and past experience, special clothing for the greenhouse worker, double-dooring entrances with air outflow from the greenhouse, and foot baths to sterilize footware before entering the greenhouse may be required. Entrance into the greenhouse should be held to a minimum, with workers being supplied the necessary facilities and tools needed for their working hours once they have entered the greenhouse.

INTEGRATED PEST MANAGEMENT

It is essential that an IPM program be developed and carefully followed to ensure that disease and insect infestations do not occur (Berlinger, 1986; Anon., 1990; Shipp et al., 1991; Clarke et al., 1994; Peet, 1996a, 1996b; Ferguson, 1996; Killebrew, 1996; Waterman, 1996; Papadopoulos et al., 1997; Snyder, 1997b). Protective and treatment procedures should be timely because after-the-fact treatment or treatments may not be able to control an established disease or insect infestation.

The insect population in the greenhouse is best monitored by placing yellow- or blue-colored sticky boards (Figure 6.6) at intervals within the plant canopy, and by examining daily the boards to determine the number and species of insects on the board (Roberts and Kania, 1996). As insect numbers accumulate on the board, procedures for insect population control can be instituted by using chemical procedures (Johnson, 1997) or by introducing predator insects into the greenhouse (Malais and Ravensberg, 1992; Gunstone, 1994; Ferguson, 1996; Kueneman, 1996).

Before any chemical or biological treatments are applied, the disease or insect species should be confirmed by a plant pathologist or entomologist, respectively, and any treatment should be applied based on procedures that conform to current chemical regulations. Greater details on disease and insect control are discussed in Chapter 8.

PLANT CULTURE SYSTEMS

The demands of the marketplace, the growing environment (such as light intensity and duration) and outside air temperatures will dictate to a considerable degree which tomato plant culture system can be efficiently employed. A single initial planting and fruit harvesting over a long period of time is one system; also there are several versions of multicropping in which the tomato plant is allowed to develop to a certain point, is topped allowing already set fruit to mature, and then is removed from the greenhouse.

For the northern Hemisphere in the upper latitudes, a two-crop system is commonly used. Planting is done in the late summer and harvesting, through December; then replanting, in March and harvesting, through the summer months, thereby avoiding the low light and low air temperature months of January and February. In the lower latitudes, plants are set in the greenhouse in late summer or early fall and then are continuously cultured and fruit harvested into the following summer months. The cycle is repeated beginning in later summer or early fall. Curry (1997) describes the system of continuous fruit production used in the greenhouses in Colorado.

There are other systems of plant culture that are being evaluated. For example, a unique automated system described by McAvoy et al. (1989a, 1989b), Giacomelli et al. (1993), and Roberts and Specca (1997) tops the tomato plant after the first

FIGURE 6.6 Sticky board hanging in tomato row for collecting insects.

cluster (truss) is set, and the plant is removed from the growing system when the fruit on that first cluster is harvested. The success of this growing system will be determined by what yield of fruit can be obtained from this one cluster. A computer simulation of this system was done by Giniger et al. (1988).

Keeping a tomato plant in profitable production over an extended period of time requires considerable management skill, while multicropping systems must be carefully managed to efficiently use the greenhouse space and maintain a constant supply of fruit to satisfy market requirements. Success depends on maximizing the greenhouse growing space for high yield fruit production, minimizing costs of production, and sustaining a flow of high quality fruit to the marketplace, thereby making the growing system, whichever is selected, conform not only to environmental conditions but also to market demands.

Mirza and Younus (1997) have summarized factors associated with greenhouse tomato production in Alberta, Canada based on a plant density of 2.7 plants per square meter, producing 63 kg/m^2 of fruit picked from March to December, with a gross revenue of $6.50/$ft^2$, and at a total operating cost of $4.80/$ft^2$. Unfortunately, similar data based on other systems of production and location are not available. The variability that exists when comparing systems is considerable, which makes comparisons difficult to make because input costs, yield, and market price are highly variable factors that impact such evaluations.

However, the rapid expansion of the greenhouse tomato industry, particularly in Canada and the United States, suggests that the various systems of growing currently being employed, particularly hydroponics (Jensen, 1997), are both profitable and able to meet consumer demands for high quality fruit. Pena (1985) describes the economic considerations for marketing and financing greenhouse vegetable production, and Savage (1989) has prepared a guide for planning a profitable hydroponic greenhouse operation.

HYDROPONIC GROWING

There are basically three hydroponic growing systems that have been or are being used to grow tomatoes commercially. Initially, the ebb-and-flow method (or modifications of the concept) was the method in wide use from the late 1930s into the 1950s. In the mid-1970s, Allan Cooper introduced his nutrient film technique (NFT), which substantially changed the basic concept of hydroponic growing; this system is relatively inexpensive to install and maintain, and is quite precise in its control of the nutrient–root environment.

With the introduction of drip irrigation combined with fertilizer injector systems, placement of water or a nutrient solution at the base of the tomato plant on a regulated basis became possible. With this type of water–nutrient solution delivery system, the use of rockwool slabs and perlite bags as the major growing media came into wide use. The introduction of the water–nutrient solution from the bottom of the rooting media is a new concept of hydroponic growing that is currently under testing for future commercial development.

Descriptions of the various hydroponic growing systems can be found in the book by Resh (1995), in the review by Parker (1994), and in the proceedings article by Rorabaugh (1995). The current state of the art of hydroponics up to 1985 can be found in the proceedings edited by Savage (1985).

EBB AND FLOW

The ebb-and-flow system consists of a growing bed (containing either gravel or sand) and a nutrient solution sump, as is illustrated in Figure 6.7. The nutrient solution is pumped periodically from the sump into the growing bed, flooding it for

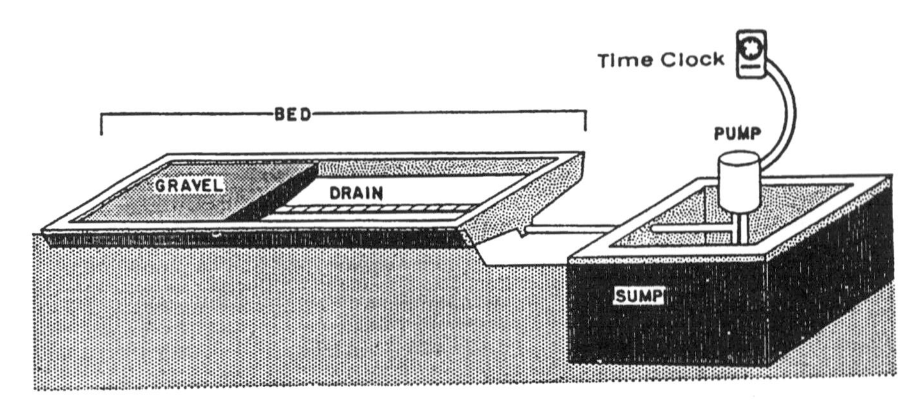

FIGURE 6.7 Ebb-and-flow hydroponic growing system.

a short period of time (5–10 min); and then the nutrient solution is allowed to drain back into the sump. The system was widely used by the U.S. Army during World War II to produce vegetables, mainly lettuce and tomato, for troops operating in the Pacific, followed by its commercial application in Florida and in various tropical regions (Eastwood, 1947). However, the ebb-and-flow system is little used today other than in hobby-type hydroponic growing systems. The method is very inefficient in its use of water and plant nutrients. Repeated use of the nutrient solution can lead to disease and nutrient imbalances, and the accumulation of precipitated substances in the gravel or sand bed, mainly calcium phosphate, will begin to significantly affect the nutrition of the plants. Therefore, periodic replacement of the growing medium is required, which adds considerably to the cost of the system over the period of its use.

Fischer et al. (1990) have described an intensive tomato production system using the ebb-and-flow technique but without a rooting medium other than a rockwool block. The tomato plant is grown in a large rockwool block that is placed on a table periodically flooded with nutrient solution. This single truss system has been described by Giacomelli et al. (1993) and Roberts and Specca (1997).

Nutrient Film Technique

A rockwool cube in which a young tomato plant has been germinated is set in a sloping trough of flowing nutrient solution. The trough usually consists of a plastic sheet that is pulled up over the cube, enclosing it in a pyramid-shaped trough as is shown in Figure 6.8. The NFT method developed by Cooper (1996), initially attracted considerable attention. However, when put into use, the technique was found to have several significant flaws that impacted its long-term use. As plant roots filled the trough, the flow of nutrient solution down the sloping trough became restricted, with the flow going either over the top of the root mass or down the sides of the plastic trough rather than through the root mass. The center of the root mass would then become anaerobic and roots would begin to die from lack of sufficient oxygen.

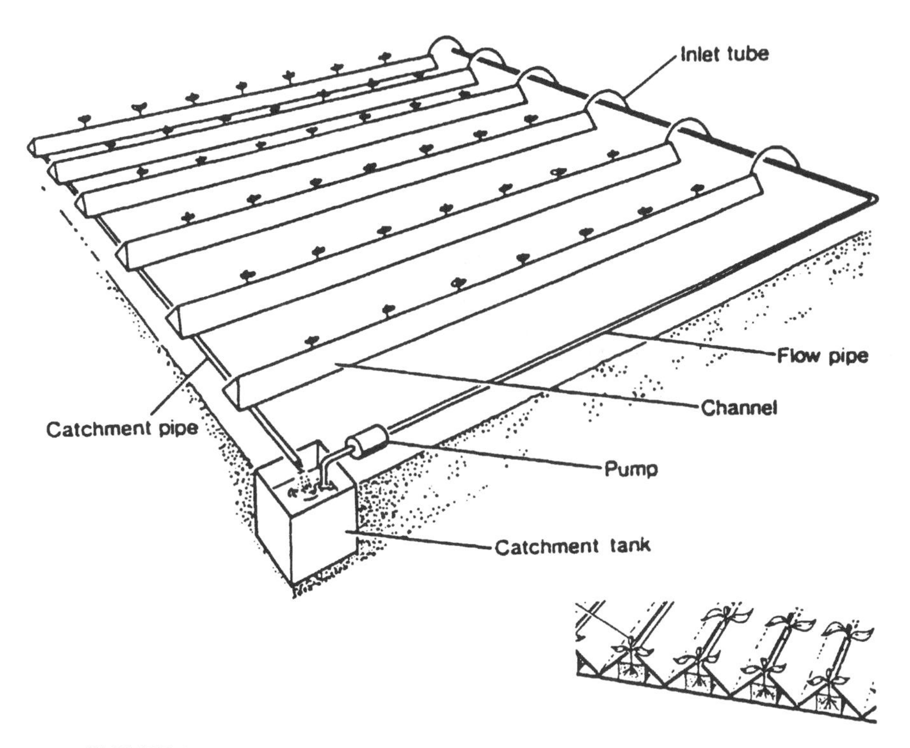

FIGURE 6.8 Nutrient flow technique (NFT) hydroponic growing system.

Altering the design of the trough to a "W" configuration, as shown in Figure 6.9, significantly changed the potential for root clogging of the channel, but the technique still presented problems in maintaining a suitable environment for best plant growth and development. The NFT concept works best for lettuce, which is a short-term crop (40–50 days) where root growth does not fill the NFT trough. The slope of the trough and the rate of nutrient solution flow down the trough can have a significant effect on the plants depending on their position in the trough, whether at the head or the foot. In addition, the NFT growing system is inefficient in its use of water and plant nutrients, and with the recirculation of the nutrient solution, disease (Evans, 1995) and nutritional problems can easily occur (Ehret and Ho, 1986).

Rockwool Slab Drip Irrigation

The rooting medium is a large (3 × 8 × 36 in.) plastic encased rockwool slab. Rockwool has excellent water-holding and aeration characteristics, making it a very desirable rooting medium (Sonneveld, 1989; Bij, 1990; Van Patten, 1991; Straver, 1996a, 1996b). This method of growing is as follows: a tomato seed is germinated in a small rockwool cube; and when the tomato seedling has initiated true leaves, the cube is placed into a larger rockwool block. Placement of the rockwool block on an opening in the rockwool slab is made when the plant roots are about to emerge from the base of the block. Block placement is shown in Figure 6.10. Small holes

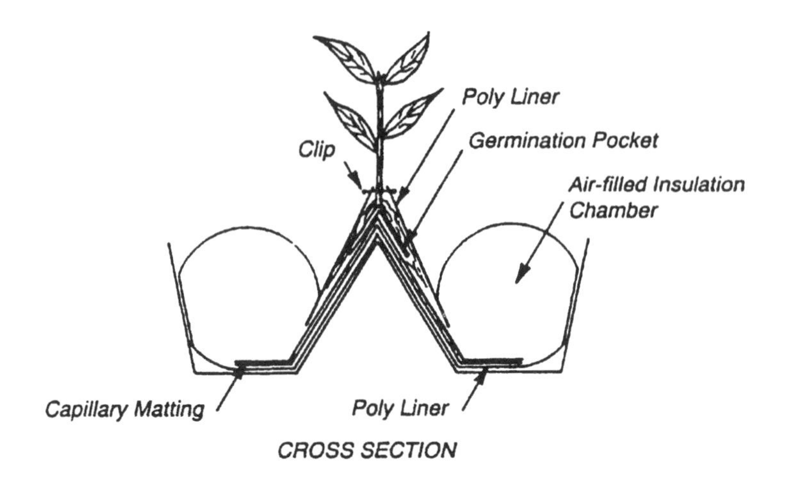

FIGURE 6.9 "W"-shaped growing trough for the NFT hydroponic growing system.

or cuts are made along the base of the plastic casing that allow excess nutrient solution to flow from the slab while keeping a small depth of nutrient solution in the bottom of the plastic casing.

The nutrient solution or water is delivered at the base of the plant on the rockwool block with sufficient flow so that the solution will flow into the rockwool slab. The management of the growing system in terms of nutrient solution composition and the frequency and amount of nutrient solution delivered to the plant is normally computer controlled (Bauerle et al., 1988; Bauerle, 1990). These controlling factors are based on environmental conditions, such as temperature and light, and plant stage of growth.

The nutrient solution that accumulates in the slab is periodically monitored for its electrical conductivity (EC); and when reaching a certain level, the slab is leached with "pure" water to remove accumulated salts, with the leaching water being applied through the drip irrigation system. Therefore, an environmentally acceptable means of disposal of the effluent from the slabs is needed. A rockwool slab can be used several times and then must be discarded. In the Colorado greenhouses, the rockwool slabs are replaced on a schedule of 16–18 months (Curry, 1997). Details on the use of this system of hydroponic tomato growing have been described by Papadopoulos (1991) and Resh (1995). For the hydroponic growing of tomato, rockwool is the most widely used rooting medium worldwide.

Perlite Bag Drip Irrigation

The rooting medium is perlite (Day, 1991) placed in a plastic bag of about the same dimensions as the rockwool slab (see preceding section). A tomato seed is germinated in either a rockwool or an Oasis® cube; and when the tomato plant has true leaves, the cube is placed into either a larger rockwool block or a cup containing either perlite or rockwool. When the roots are about to emerge from the base of the block or cup, the plant is placed into an opening in the perlite bag, as has been described by Brentlinger (1992), Gerhart and Gerhart (1992), and Resh (1995).

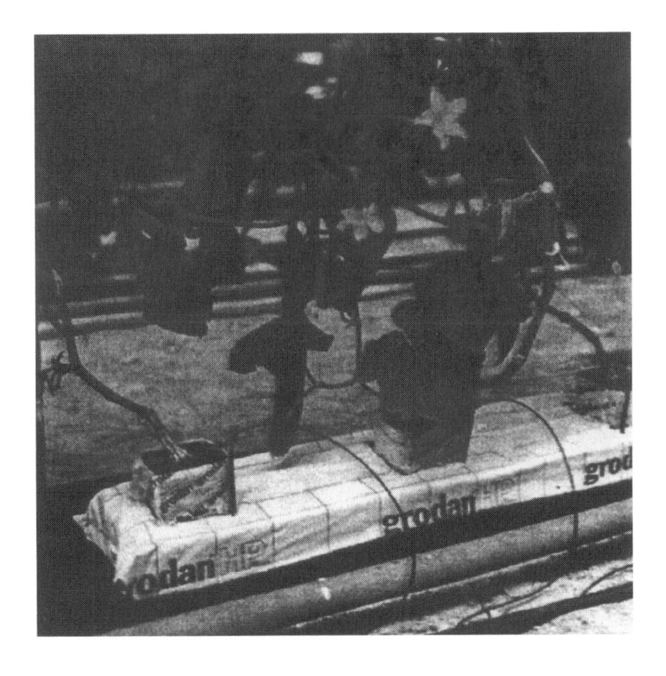

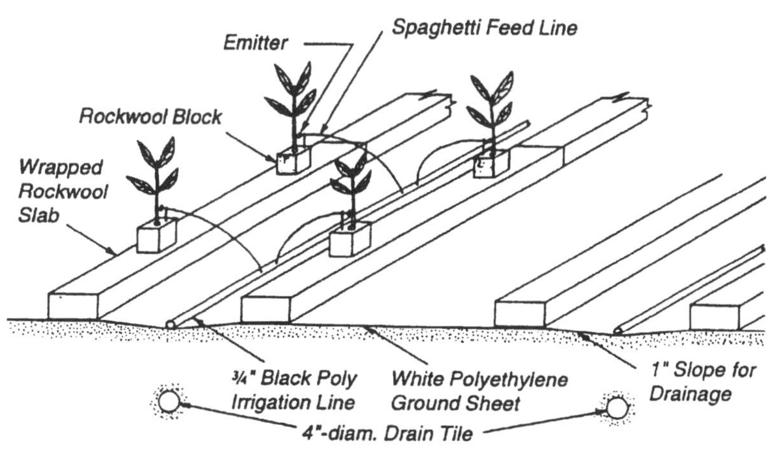

FIGURE 6.10 Rockwool slab hydroponic growing system.

The nutrient solution or water is delivered to the base of the plant in the rockwool block or cup by means of a drip irrigation system (Figure 6.11). The composition of the nutrient solution and its schedule for delivery are based on environmental conditions and plant growth stage as was described for the rockwool slab method. Small holes or cuts are made along the base of the plastic bag that allow excess nutrient solution to flow from the bag while keeping a small depth of nutrient solution in the bottom of the bag. The nutrient solution in the perlite bag is monitored for its EC; and when the EC reaches a certain level, the bag is leached with pure water applied through the drip irrigation system. Therefore, an environmentally acceptable

FIGURE 6.11 Perlite bag hydroponic growing system.

means of disposal of the effluent from the perlite bags is needed. The perlite in the bag can be used to produce two crops and then must be discarded.

A detailed description of this hydroponic method of tomato production can be obtained from CropKing (5050 Greenwick Road, Seville, OH 44273). This system of growing tomatoes hydroponically is in fairly wide use, mainly in the United States.

AquaNutrient Growing System

With the AquaNutrient system, introduced by the author (Jones, 1997c), plants are grown in a trough, pot, or pot–trough system in which a constant level of nutrient solution is maintained in the bottom of the growing vessel. The rooting medium is either perlite or a mixture of perlite and pine bark. The system has yet to be put into commercial use on a large scale. The system is very efficient in its use of water and plant nutrients since all the water and nutrients supplied are utilized by the plant. In addition, adjustments of the nutrient solution supplied to the plants based either on growing conditions or on stage of plant growth are not required. The concept for the AquaNutrient System is based on the "quantity and balance of nutrients" concept first described and developed by Geraldson (1963, 1982) for stake tomato production on the sandy soils in central Florida, and the design of the system is illustrated in Figure 6.12. The initial concept of the subirrigation procedure for both greenhouse and small container use has been described by Bruce et al. (1980) and Jones (1980), respectively.

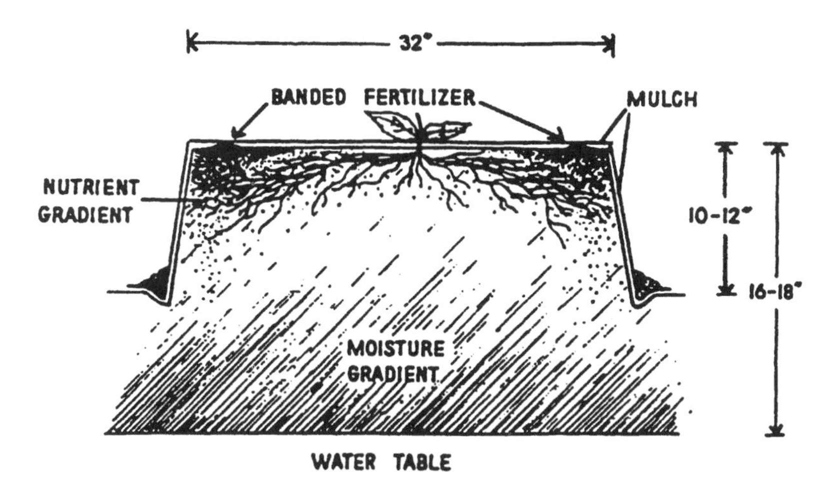

FIGURE 6.12 Control of the root ionic environment obtained by handling fertilizer on the surface of a raised bed, using a plastic mulch cover and maintaining a definite water table.

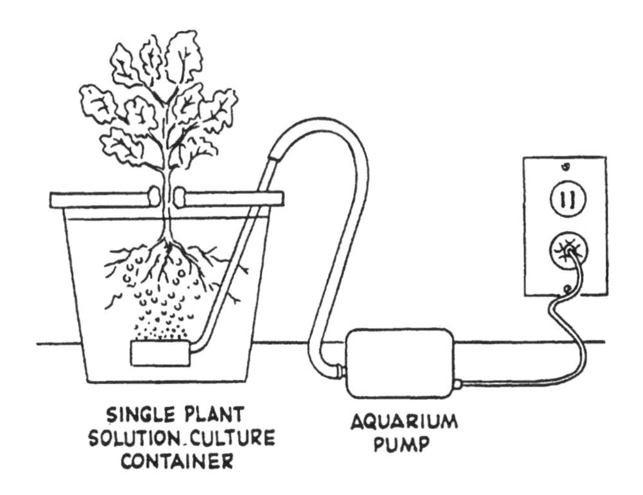

FIGURE 6.13 Standing-aerated hydroponic growing system.

Standing-Aerated

In this hydroponic growing system, the plant roots are suspended in a nutrient solution that is being continuously aerated (Figure 6.13); this system of plant growing is primarily used for plant nutrition studies since the composition of the nutrient solution can be easily manipulated. However, the standing-aerated system is not suitable for large-scale commercial production of plants.

Aeroponics

In an aeroponic system, the plant roots are suspended in a fine mist of nutrient solution that is applied on a continuous or intermittent basis. Aeroponic growing systems have been described by Soffer (1985, 1988), and commercial details on the method has been given by Adi Limited (1982); however, the aeroponic technique has yet to be found economically suitable for the large-scale production of plants.

THE NUTRIENT SOLUTION

Of the 16 essential elements required by plants, 13—nitrogen (N), phosphorus (P), potassium (K), calcium (Ca), magnesium (Mg), and sulfur (S), known as the *major elements;* and boron (B), chlorine (Cl), copper (Cu), iron (Fe), manganese (Mn), molybdenum (Mo), and zinc (Zn), known as the *micronutrients*—must be present at specific concentrations in the nutrient solution supplied to the plant to sustain normal growth (Jones, 1997a). The composition of nutrient solutions and their use with various hydroponic or soilless growing systems have been reviewed by Jones (1997b). Techniques for preparing and maintaining a nutrient solution have been discussed by Gerber (1985), Sonneveld (1985), Berry (1989), Muckle (1990), Wilcox (1991), Schon (1992), and Bugbee (1995).

Most hydroponic nutrient solution formulations are based on Hoagland's original formula (Hoagland and Arnon, 1950), as given in Table 6.5.

The composition of nutrient solution will varying depending on the formulation used, but the ranges in composition for a typical nutrient solution and ionic forms are given in Table 6.6.

The nutrient solution when initially composed should be analyzed by a competent laboratory (see page 70 for laboratory sources) to ensure that all the elements are present at their desired concentration. If an injection system is being used to dispense a nutrient solution concentrate, the composition of the nutrient solution being delivered through the drip irrigation system should also be monitored (assayed) periodically for the same purpose of ensuring proper concentration delivery.

The composition of a nutrient solution is frequently adjusted based on the method of growing, stage of plant growth (Ward, 1964; Sonneveld, 1985; Bloom, 1987; Wilcox, 1991; Voogt and Sonneveld, 1997), and changing environmental conditions. The frequency of nutrient solution delivery by drip irrigation, or time and rate of flow for other growing procedures, is standard practice and has been discussed in general by Jones (1997b); more specifically by Schippers (1979), Ames and Johnson (1986), Molyneux (1988), Papadopoulos (1991), Cooper (1996); by Hochmuth and Hochmuth (1996), for the NFT growing system; and by Papadopoulos (1991), Straver (1996a, 1996b), and Hochmuth and Hochmuth (1996), for the rockwool slab–drip irrigation growing system.

In both the rockwool slab–drip irrigation and perlite bag–drip irrigation systems, the solution in the slab or bag must be monitored due to the accumulation of salts that occurs with time in the slab or perlite, an accumulation that can eventually reduce plant growth. The procedure is to draw an aliquot of the solution from the slab or bag and measure its EC. When the EC exceeds a predetermined level, the

TABLE 6.5
Hoagland's Nutrient Solution Formulas

Stock Solution	To use mL L^{-1}
Solution No. 1	
1*M* potassium dihydrogen phosphate (KH_2PO_4)	1.0
1*M* potassium nitrate (KNO_3)	5.0
1*M* calcium nitrate [$Ca(NO_3)_2.4H_2O$]	5.0
1*M* magnesium sulfate ($MgSO_4 \cdot 7H_2O$)	2.0
Solution No. 2	
1*M* ammonium dihydrogen phosphate ($NH_4H_2PO_4$)	1.0
1*M* potassium nitrate (KNO_3)	5.0
1*M* calcium nitrate [$Ca(NO_3)_2 \cdot 4H_2O$]	5.0
1*M* magnesium sulfate ($MgSO_4 \cdot 7H_2O$)	2.0
Micronutrient Stock Solution	**g L^{-1}**
Boric acid (H_3BO_3)	2.86
Manganese chloride ($MnCl_2 \cdot 4H_2O$)	1.81
Zinc sulfate ($ZnSO_4 \cdot 5H_2O$)	0.22
Copper sulfate ($CuSO_4 \cdot 5H_2O$)	0.08
Molybdate acid ($H_2MoO_4 \cdot H_2O$)	0.02
Iron	
For Solution No. 1: 0.5% iron ammonium citrate	To use 1 mL L^{-1}
For Solution No. 2: 0.5% iron chelate	To use 2 mL L^{-1}

slab or bag is then leached with pure water through the drip irrigation system. The optimum and acceptable range in element composition in a rockwool slab for tomato has been established by Ingratta et al. (1985), as given in Table 6.7.

When the solution in the rockwool slab is above the "acceptable range," the rockwool is leached with pure water to remove the accumulated salts.

NUTRIENT SOLUTION MANAGEMENT

The challenge for the hydroponic grower is to maintain the nutrient element status of the tomato plant to keep it productive over an extended period of time. The initial composition of the nutrient solution, its rate of delivery, and adjustment in composition with both the changing status of the plant and environmental conditions are significant factors (Bloom, 1987; Berry, 1989; Schon, 1992; Bugbee, 1995). It is not possible to cover all these aspects in this discussion, but some guidelines can be given.

Nutrient Film Technique

The theoretically ideal nutrient solution formula for the NFT system given by Cooper (1996) is shown in Table 6.8.

This formula gives the following essential element concentrations in the "starter" solution:

Element	Concentration [mg L^{-1} ppm]
Major Elements	
Nitrogen (N)	200
Phosphorus (P)	60
Potassium (K)	300
Calcium (Ca)	170
Magnesium (Mg)	50
Micronutrients	
Boron (B)	0.3
Copper (Cu)	0.1
Iron (Fe)	12.0
Manganese (Mn)	2.0
Molybdenum (Mo)	0.2
Zinc (Zn)	0.1

When additional nutrient solution is needed to replace what has been absorbed by the plants, a "topping-up solution" is added to the "starting solution" of the following composition:

Element	Concentration [mg L^{-1} ppm]
Major Elements	
Nitrogen (N)	140
Phosphorus (P)	—
Potassium (K)	147
Calcium (Ca)	180
Magnesium (Mg)	32
Micronutrients	
Boron (B)	0.32
Copper (Cu)	0.065
Iron (Fe)	1.5
Manganese (Mn)	0.3
Molybdenum (Mo)	0.007
Zinc (Zn)	0.1

With the NFT system, one method for determining when to make a nutrient solution irrigation is that for every 0.3 MJ m^{-2} of radiation received, an irrigation is scheduled. Similar procedures are being used for other hydroponic systems, regulating the frequency and amount of nutrient solution applied based on solar radiation received.

TABLE 6.6
Major and Micronutrient Ionic Forms and Normal Concentration Ranges in the Nutrient Solution

Element	Ionic Form	Concentration in Solution [mg L^{-1} (ppm)]
	Major Elements	
Nitrogen (N)	NO_3^- or NH_4^+	100–200
Phosphorus (P)	HPO_4^{2-} or $H_2PO_4^-$ [a]	30–50
Potassium (K)	K^+	100–200
Calcium (Ca)	Ca^{2+}	100–200
Magnesium (Mg)	Mg^{2+}	30–70
	Micronutrients	
Boron (B)	BO_3^{3-} or H_3BO_3 [b]	0.2–0.4
Chloride (Cl)	Cl^-	5.0
Copper (Cu)	Cu^{2+}	0.01–0.1
Iron (Fe)	Fe^{2+} or Fe^{3+}	2–12
Manganese (Mn)	Mn^{2+}	0.5–2.0
Molybdenum (Mo)	MoO_4^{2-}	0.05–0.2
Zinc (Zn)	Zn^{2+}	0.05–0.10

[a] The form depends on the pH of the nutrient solution.
[b] It is being increasingly suggested that boron exists in the nutrient solution as molecular H_3BO_3.

The influence of stage of plant growth is also a factor in determining what the elemental concentration ranges should be, as has been suggested by Hochmuth and Hochmuth (1996) for the NFT and rockwool techniques (Table 6.9). As the stage of growth advances, there is an increase in the N, K, and Mg concentrations, while the other elements remain at constant concentration.

Other Hydroponic Systems

Research is needed to determine how best to maintain a nutrient solution to ensure nutrient element sufficiency for the tomato crop as it advances through each stage of its growth cycle. This is particularly true if the nutrient solution is recirculated. The composition of the nutrient solution is commonly adjusted based on stage of growth and climatic conditions, mainly total solar radiation received, adjusting primarily the major elements, N and K. The concentration of each element in the nutrient solution depends considerably on the relationship among the number of plants, volume of solution applied per plant, and frequency of application (Jones, 1997b). Jones (1997) has suggested that a constant level of nutrient element supply is required for best plant growth, the composition of the nutrient solution being dilute in comparison with that commonly found in Hoagland (see Table 6.5) or modified Hoagland solutions. There is still much to be learned about how best to

TABLE 6.7
Optimum Concentrations and Acceptable Ranges of Nutrient Solution in the Substrate

Determination	Optimum	Acceptable Range
EC (dS m^{-1})	2.5	2.0–3.0
pH	5.5	5.0–6.0
	mg L^{-1} (ppm)	**mg L^{-1} (ppm)**
Bicarbonate (HCO_3)	<60	0.60
Nitrate (NO_3)	560	370–930
Ammonium (NH_4)	<10	0–10
Phosphorus (P)	30	15–45
Potassium (K)	200	160–270
Calcium (Ca)	200	160–280
Magnesium (Mg)	50	25–70
Sulfate (SO_4)	200	100–500
Boron (B)	0.4	0.2–0.8
Copper (Cu)	0.04	0.02–0.1
Iron (Fe)	0.8	0.4–1.1
Manganese (Mn)	0.4	0.2–0.8
Zinc (Zn)	0.3	0.2–0.7

supply the growing plant with its nutrient element needs in order to maintain sufficiency during the entire growth cycle of the plant.

Filtering and Sterilization

If a nutrient solution is recirculated, in addition to its composition being maintained, any accumulated organic material from the plant roots must be removed by filtering, and the solution must be sterilized. Cartridge-type swimming pool filters will remove most suspended materials and Millipore® filtering will remove large molecular substances, providing a degree of disease control. Evans (1995) has suggested various procedures for sterilization, one being the use of ultraviolet (UV) lamps as recommended by Buyanovsky et al. (1981). Two 16-W UV lamps are placed in the path of the flowing nutrient solution flowing at 13 L (3 gal) min^{-1}.

ORGANIC MEDIA BAG CULTURE SYSTEMS

Prior to the current accepted use of several forms of hydroponics, growing in soil as described by Wittwer and Honma (1969) and Brooks (1969) was the commonly used growth medium. The next innovation was bag culture (Sheldrake, 1980; Bauerle, 1984), using some form of "Cornell Peat-lite" mixes (Boodley and Sheldrake, 1972), or customized mixes containing various substances that control both the physical and chemical characteristics of the mix (Bunt, 1988).

TABLE 6.8
Nutrient Solution Formula to Give the Theoretically Ideal Concentration of Essential Elements

Reagent	Formula	Amount (g 1000 L^{-1})
Potassium dihydrogen phosphate	KH_2PO_4	263
Potassium nitrate	KNO_3	583
Calcium nitrate	$Ca(NO_3)_2 \cdot 4H_2O$	1003
Magnesium sulfate	$MgSO_4 \cdot 7H_2O$	513
EDTA iron	$[(CH_2\text{-}N(CH_2\text{-}COOH)_2]FeNa$	79
Manganese sulfate	$MnSO_4 \cdot H_2O$	6.1
Boric acid	H_3BO_3	1.7
Copper sulfate	$CuSO_4 \cdot 5H_2O$	0.39
Ammonium molybdate	$(NH_4)_6Mo_7O_{24} \cdot 4H_2O$	0.37
Zinc sulfate	$ZnSO_4 \cdot 7H_2O$	0.33

TABLE 6.9
Final Delivered Nutrient Solution Concentration for Hydroponic (NFT-PVC Pipe and Rockwool) Tomato in Florida Greenhouses

	Stage of Growth [mg L^{-1}, ppm]				
Element	1	2	3	4	5
Major Elements					
Nitrogen (N)	70	80	100	120	150
Phosphorus (P)	50	50	50	50	50
Potassium (K)	120	120	150	150	150
Calcium (Ca)	150	150	150	150	150
Magnesium (Mg)	50	50	50	60	60
Micronutrients					
Boron (B)	0.7	0.7	0.7	0.7	0.7
Copper (Cu)	0.2	0.2	0.2	0.2	0.2
Iron (Fe)	2.8	2.8	2.8	2.8	2.8
Manganese (Mn)	0.05	0.05	0.05	0.05	0.05
Zinc (Zn)	0.3	0.3	0.3	0.3	0.3

Note: Stage 1, transplant to first cluster; stage 2, first cluster to second cluster; stage 3, second cluster to third cluster; stage 4, third cluster to fifth cluster; stage 5, fifth cluster to termination.

Papadopoulos (1991) has described several organic mix formulations for growing tomatoes: equal ratios for peat moss and horticultural vermiculite; or an equal mix of peat moss, horticultural vermiculite, and perlite placed in a 42-L plastic bag

TABLE 6.10
Ingredients for a Complete and Case Mixture of Peat Moss and Vermiculite (1 m³)

Ingredient	Complete	Base
Peat moss	0.5 m^3	0.5 m^3
Horticultural vermiculite	0.5 m^3	0.5 m^3
Ground limestone (dolomite)	7.5 kg	—
Limestone (pulverized FF)	—	5.9
Gypsum (calcium sulfate)	3.0 kg	—
Calcium nitrate	0.9 kg	—
Superphosphate, 20%	1.5 kg	1.2 kg
Epsom salts (magnesium sulfate)	0.3 kg	0.3 kg
Osmocote 18–6–12 (9 months)	5–6 kg	—
Chelated iron (NaFe 138 or 330), 10%	30 g	35 g
Fritted trace elements (FTE 503)	225 g	110 g (or FTE 302)
Borax (sodium borate)	—	35 g

that measures 35 × 105 cm when flat. Dolomitic limestone and various fertilizers (i.e., superphosphate, potassium nitrate, calcium nitrate, magnesium sulfate, and micronutrients) are added to the mixes to supply the needed essential elements. Papadopoulos (1991) presents two concepts: one in which all the required nutrients are added to the mix initially, or the other in which a portion is added and then liquid fertilizer is added to the irrigation water as needed during the growing season. The compositions of the two mixes are given in Table 6.10.

Bunt (1988) gave the formula for a tomato bag mix of peat nodules (sedge or humified sphagnum) in 20-L bag as follows:

Ingredient	kg m^{-3}	yd^3
Superphosphate (0–20–0)	1.75	3 lb
Potassium nitrate	0.87	1 lb 8 oz
Potassium sulfate	0.44	12 oz
Ground limestone	4.2	7 lb
Dolomitic limestone	3.0	5 lb
Frit 253A	0.4	10 oz

Note: Additional slow-release nitrogen as 0.44 kg m^{-3} urea–formaldehyde (167 mg N L^{-1}) is sometimes included. If slow-release phosphorus fertilizer is required, magnesium ammonium phosphate ("MagAmp" or "Enmag") at 1.5 kg m^{-3} is added.

Another bag organic mix suitable for tomato culture consists of the following ingredients:

- Sphagnum peat moss—9 bushels
- Vermiculite—9 bushels

- Perlite—4 bushels
- Dolomitic limestone—8 lb
- Superphosphate fertilizer (0–20–0)—2 lb
- Calcium nitrate—1 lb
- Borax—10 g
- Chelated iron—35 g

Bruce et al. (1980) grew greenhouse tomatoes in pure milled pine bark with good success in a growbox system that was suitable for home garden use as described by Jones (1980). All the nutrient elements required by the tomato plant were put into the organic mix. Various formulations of soilless organic mixes have been described by Jones (1997b).

A simple formulation would be a mix consisting of:

- Milled pine bark—9 bushels
- Dolomitic limestone—1 lb
- Fertilizer 10–10–10—1 lb

However, with extended use, additional fertilizer will have to be added to maintain a tomato plant in good nutritional status.

Keeping the water pH of an organic mix less than 5.5 is essential to prevent possible micronutrient deficiencies from occurring, mainly B, Mn, and Zn, because the availability of these elements begins to drop sharply with increasing pH. If "hard water" is used for irrigation, less limestone (dolomitic or otherwise) should be initially added to the mix because there may be sufficient Ca and possibly Mg in the water to satisfy the crop requirement (see pages 97–98).

Although some growers may still be using an organic mix, there is little interest today in developing modifications of these mixes for improved performance under greenhouse-growing conditions for the production of tomatoes.

A major requirement for bag culture is control of watering to ensure sufficiency but not excess. Adams (1990) compared several watering regimes for a peat bag system, finding that restricting water to 80% of the estimated requirement resulted in a 4% loss in yield (smaller fruit) but improved flavor components. Adams (1990) also found that high water levels resulted in Mn deficiency. Snyder and Bauerle (1985) found that bag size influenced fruit yield and the incidence of BER, 7-L bags producing lower yields and high incidence of BER, while 14-, 21-, and 35-L bags produced about the same yield and lower BER.

7 Seed and Seedling Production

CONTENTS

SEED CHARACTERISTICS

The tomato seed is 3–5 mm in size, silky in appearance, flat, and light cream to brown in color. It contains a large coiled embryo surrounded by a small amount of endosperm (Figure 7.1).

The weight of an individual seed varies considerably, with 300–350 seeds weighing 1 gram. Put another way, there are 7,000–12,000 tomato seeds per ounce of seed. Based on one source, an ounce of seed would be needed to produce 4,000 plants.

Approximate seed count by weight is

	Number of Seeds	
Ounce	**Small–Large Fruit**	**Cherry**
1/32	250–375	375
1/16	500–750	750
1/8	1000–1500	1500
1/4	2000–3000	3000

Tomato seeds are not mature, and therefore viable, until the tomato fruit is mature. For the home gardener, seed recovered from mature fruit may not come true to the variety due to cross pollination. Therefore, for best results, seed should be obtained from a reliable seed supplier.

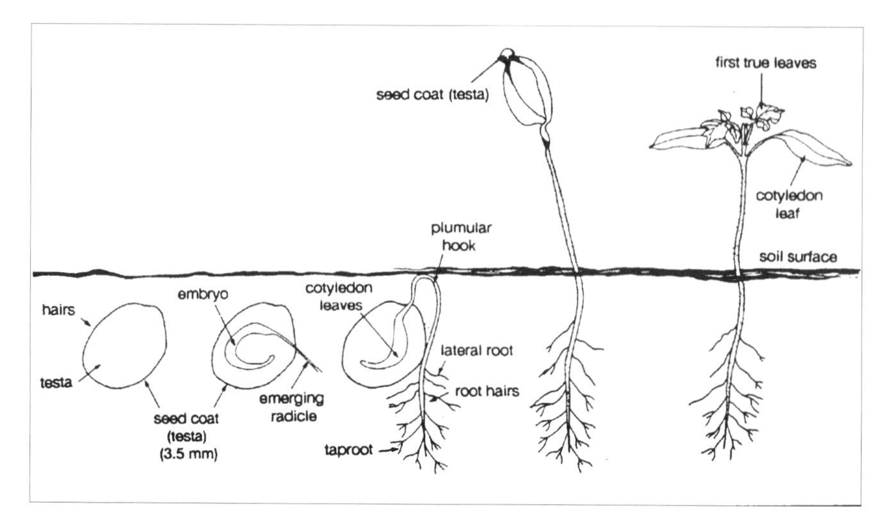

FIGURE 7.1 Tomato seed components and process of germination to produce tomato plant seedling. (Reproduced from Papadopoulos, A.P. 1991. *Growing Greenhouse Tomatoes in Soil and in Soilless Media*. Agriculture Canada Publication 186/E. Communications Branch, Agriculture Canada, Ottawa, Canada. Reproduced with the permission of the Minister of Public Works and Government Services Canada 1998.)

Mature seed can remain viable for up to 4 years in hermetically sealed containers at a seed moisture content of 5.5%. The estimated maximum safe seed moisture content for 1 year storage at different temperatures is

Temperature [°F (°C)]	Moisture (%)
40–50 (4–10)	13
70 (21)	11
80 (26)	9

For the production of seed in the United States, yield of tomato seeds is on the average 121 lb A^{-1}, with very good seed yields being as high as 200 lb A^{-1} (Maynard and Hochmuth, 1997).

GERMINATION TESTING AND LABELING

The test procedures for determining tomato seed germination percentage is based on procedures published in the United States *Federal Register*, volume 59, no. 239, December 14, 1994. The seed is placed between blotters; in a Petri dish covered with two layers of blotters with one layer of absorbent cotton, with five layers of paper toweling, with three thicknesses of filter paper, or with sand or soil; or on blotters, with the temperature ranging from 68 to 86°F (20 to 30°C). The seed is treated with a solution of potassium nitrate (KNO_3), with germination taking place in the light. First germination counts are made at 5 days and then at 14 days. The results from such a germination test are then placed on the seed label. The seed label must include

- Kind, variety, and hybrid
- Name of shipper or consignee
- Germination
- Lot number
- Seed treatment

GERMINATION FOR PLANT PRODUCTION

The minimum temperature for tomato seed germination is 46.6–50°F (8–10°C), the range in temperature for varying degrees of germination being:

Characteristic	Seed Germination Temperature [°F (°C)]
Minimum	50 (10)
Optimum range	60–85 (16–29.5)
Optimum	85 (29.5)
Maximum	95 (35)

The days required for seedling emergence vary with temperature, as is seen in the following:

Soil Temperatures for Germination [°F (°C)]	Seedling Emergence (days)
50 (10)	43
59 (15)	14
68 (20)	8
77 (25)	6
86 (30)	6
95 (35)	9

Tomato seeds germinate best in the dark, although there are some tomato cultivars that will not germinate in the light.

YOUNG VEGETATIVE PLANT

On seed germination, a taproot emerges and the stem elongates with the cotyledons encased in the seed coat. The seed coat falls away, and the cotyledons emerge horizontal with the stem (hypocotyl). In several days, the first "true leaves" emerge and the root system begins to branch, producing a fibrous roots with adventitious roots emerging from the lower stem. A young vegetative seedling is shown in Figure 7.2. A detailed description of seed germination and the developing seedling from seed to full maturity is given by Picken et al. (1986).

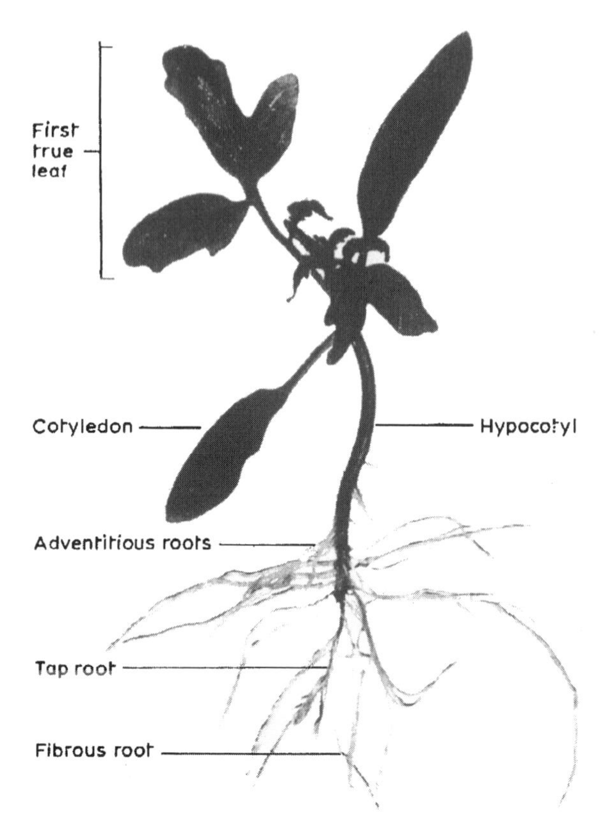

FIGURE 7.2 Young tomato plant seedling components. (From Picken, A.J.F., K. Stewart, and D. Kalpwijk. 1986. p. 207. In: J.G. Atherton and J. Rudich (Eds.), *The Tomato Crop: A Scientific Basis for Improvement.* Chapman & Hall, NY. With permission.)

SEEDLING PRODUCTION

Planting seeds directly into the garden or field soil is not recommended because the cost for seed, time required to prepare and manage the seedbed, and frequently low germination make the procedure impractical and costly. Therefore, the production of a seedling in a controlled environment for transplanting into the garden or field is the procedure recommended and normally used.

The production of a seedling, frequently referred to as a *transplant*, is an extremely important procedure because future plant growth and fruit production is affected by the character of the seedling produced. Snyder (1995) has described procedures for producing a tomato transplant, detailing the selection of the growing media, seeding technique, fertilization, temperature, watering, and light conditions, all important factors that will affect the quality of the transplant produced.

Vavrina and Orzolek (1993) have written a review article on tomato transplant production covering the publication period from 1927 to 1991, focusing on the age factor influencing future fruit yields. In agreement with Leskovar and Cantliffe (1990), transplants that are between 3 and 5 weeks old are considered "ideal," while transplants over 5 weeks old are less desirable. However, it has been observed that the younger transplants will produce more fruit, while older transplants will produce less fruit but larger in size. For example, transplants over 9 weeks old were found to produce extralarge fruit. Vavrina (1991) also discussed the effect that transplant age makes on the final performance of the plant.

The storage temperature for transplants awaiting planting for a period of over 2 days should be below ambient temperature, and at 41–59°F (5–15°C) for storage over longer periods of time. Even at temperatures between 50 and 55.4°F (10 and 13°C), Leskovar and Cantliffe (1990) found that root growth was not inhibited. Growth and fruit yields for 35-day-old transplants were more affected by transplant handling than older transplants were; therefore, greater care is needed to minimize transplant stress.

An important process for preparing the seedling for transplanting into the field or garden is "hardening," a procedure that will acclimate the seedling prior to its placement in the outside environment. Hardening normally takes about 2 weeks, with the seedling being placed in a cool and shady environment during this period. In addition, watering is reduced and no fertilizer is given to the plant, because the objective is to slow or stop additional height growth, although the plant itself may become larger in girth size and roots may further expand.

Another procedure that will reduce the elongation of the seedling is "brushing," a daily gently brushing of the top of the plant with a soft brush, or the same effect can be obtained by slowly and gently moving a smooth stick just touching the top of the plant. This procedure is widely used for the commercial production of transplants for processing tomatoes, transplants that are grown in Georgia, for example, and then shipped and planted in fields in Indiana, Michigan, and Ohio.

For those wanting to "do it on their own," Meyer (1998) has described a procedure that can be used by the home gardener. His procedure requires the following basic supplies (the first four items on this list are usually included with a purchased complete seedling tray; a complete starting seedling tray, Figure 7.3, can be purchased from most garden supply stores with all the ingredients needed included other than seed):

Cell pack: plastic tray with "cells" or 3-in. deep pockets with drainage holes
Solid plastic tray: sufficient size to accommodate the cell pack to water from the bottom
Seed-starting mix: standard seed starting mix of a blend of peat moss, perlite, or vermiculite
Plastic cover: sufficient size to cover the cell pack tray
Lights: ordinary fluorescent "shop lights," with new bulbs
Liquid fertilizer: use of a formulation that is for seedlings, an organically based fertilizer is being recommended
Pots: peat or paper pots for receiving the developing seedling when of sufficient size

FIGURE 7.3 Commercial tray suitable for producing tomato seedlings.

If a purchased seedling tray system is used, the growth medium may contain sufficient added fertilizer to ensure good initial growth. Therefore, the instructions provided with the seedling tray need to be carefully read and followed.

For the commercial grower as well as the home gardener, the use of rockwool or Oasis® cubes (Figure 7.4) as the starting medium are commonly used. Some Oasis® cubes contain added fertilizer so that none needs to be added during initial seedling growth. For the use of rockwool cubes, a small amount of fertilizer should be added to the water. The cubes should be placed in a plastic tray on a flat surface; and water or a dilute nutrient solution [the author uses one-fifth Hoagland's solution (see Table 6.5) for producing tomato seedlings] should be carefully poured into the tray, allowing the cube to wet from the bottom to the top, only applying sufficient

FIGURE 7.4 Oasis® (left) and rockwool (right) seedling cubes.

water to moisten the cube with no excess water left in the tray. These cubes normally have a small hole in the top of the cube where the tomato seed is placed. A clear plastic cover is then placed over the tray. No additional water or nutrient solution should be necessary until the seedlings begin to produce true leaves (see Figure 7.2). The cubes can be checked periodically by slightly pinching the cube that should be moist to the feel. If not moist, additional water or nutrient solution can be added as before, sufficient to moisten the cube with no excess left in the tray.

Normally rockwool and Oasis® cubes come as sheets so that many plants can be seeded in one sheet, making handling easy. The sheets are scored so that each individual cube can be separated from the sheet. The cubes containing the seedling can be easily separated and placed into a larger rockwool block or growth medium-containing vessel. Therefore, the seedling is easily moved with minimum disturbance of the roots.

Tomato seeds are quite small and somewhat difficult to handle. By placing a few seeds in a small spoon or on a small trowel, a small pick can be used to push a seed off the edge into the cell or hole in the cube. One or two seeds can placed into each cell or hole in the cube. The quality of seed today is such that high (75%+) germination percentage (normally stated on the seed container) is expected. Therefore, if two seeds are placed in each cell or cube, and both seeds germinate, then one of the seedlings will have to be removed (not an easy task if a large number of seedlings are being produced). Therefore, one seed per cell or cube would normally be sufficient.

Place the covered seedling tray in a warm place, 70°F (21°C) being best, until all the seeds germinate. The initial leaves are called cotyledons to be followed by the development of two initial true leaves 4–6 days later. Once the seeds germinate, the tray lid is removed and the tray is placed in the light, either in natural sunlight if sufficient in terms of intensity and duration or under artificial lights using the fluorescent lights to give a 16-h day of light. The fluorescent bulbs should be placed just above the seedlings. The best air temperature for the seedlings at this stage is between 60 and 65°F (16 and 18.5°C).

The growth medium should be watered as needed to keep it moist, avoiding overwatering that can slow growth and promote conditions that could lead to root disease development. Maintaining light air movement through the seedling canopy will keep the seedlings dry and reduce the potential for fungus disease development.

When the first true leaves emerge and there are two plants in the cell or cube, remove one of the seedlings by cutting the unwanted plant at the growing medium surface.

Depending on the growing conditions and use of the plants, they should be transplanted when the roots begin to grow out of the cubes or out of the bottom of the cell. If the plants are to be placed outdoors in a garden, the seedlings should be hardened by keeping the plants at 60°F (15.5°C) in a shady environment for about 2 weeks. A healthy looking transplant is shown in Figure 7.5.

TRANSPLANTS

For the home gardener in the United States, transplants are readily available in the spring from garden stores and centers and even in some grocery stores. The gardener needs to carefully select these plants, being sure they are free of any pests and that the color and shape of the plant indicates that they have been correctly produced and are being properly maintained. A poorly produced and maintained transplant can be very slow in becoming acclimated to the environment into which it is planted. Although today most transplants for home garden use have been properly produced, they are, too frequently and unfortunately, not properly maintained by the retailer. For the home gardener, it would be desirable to purchase transplants at the time they first arrive at the garden store or center, or if possible, obtain them directly from the producer.

Today, several seed companies also have transplants available (see seed sources in the next section) that can be obtained through the mail or other package delivery services.

Transplants should not be placed in the soil unless the soil temperature is above 60°F (15.5°C), although the tomato seedling will survive at a lower soil temperature. Tomato roots will not grow vigorously until the soil temperature is 70°F (21°C) or higher. If the air temperature drops below 43°F (6°C), the young seedlings should be covered. Today, there are some varieties that can withstand cool air and soil temperatures; however most of the more common varieties will not tolerate low air and soil temperatures. Hessayon (1997) has described techniques for the successful home production of tomato as well as other common garden vegetables. In addition, information on tomato production can be found in other garden books and gardening magazines.

FIGURE 7.5 Healthy tomato transplant.

SEED SOURCES

The following is a list of some of the major suppliers of tomato seeds:

Abundant Life Seed Foundation, P.O. Box 772, Port Townsend, WA 98368
Agrisales, Inc., P.O. Box 2060, Plant City, FL 33564
De Ruiter Seed, Inc., P.O. Box 20228, Columbus, OH 43220
Garden State Heirloom Seed Society, P.O. Box 15, Delaware, NJ 07853

Gurney's Seed & Nursery Company, 110 Capital Street, Yankton, SD 57079
Heirloom Seed Project, Landis Valley Museum, 2451 Kissel Hill Road, Lancaster, PA 17601
Heirloom Seeds, P.O. Box 245, West Elizabeth, PA 15808
Johnny's Selected Seeds, 310 Foss Hill Road, Albion, ME 04910
Maine Seed Saving Network, P.O. Box 126, Penobscot, ME 04476
Park Seed Company, 1 Parkton Avenue, Greenwood, SC 29647-0001
Seed Savers Exchange, 3076 North Winn Road, Decorah, IA 52101
Seeds of Diversity Canada, P.O. Box 36, Station Q, Toronto, Ontario, Canada M4T 2L7
Southern Exposure Seed Exchange, P.O. Box 170, Earlysville, VA 22936
Sunseeds, 18640 Sutter Boulevard, Morgan Hill, CA 95037-2825
Terra Edibles, Box 63, Thomasburg, Ontario, Canada K0K 3H0
Territorial Seeds Ltd., P.O. Box 157, Cottage Grove, OR 97424
Tomato Growers Supply Company, P.O. Box 2237, Fort Myers, FL 33902
Totally Tomatoes, P.O. Box 1626, Augusta, GA 30903
Turtle Tree Seed Farm, 5569 N. County Road 29, Loveland, CO 80538
Vegetable Seed Sources (Donald N. Maynard), GCREC-Bradenton Extension Report BRA-1996-1, University of Florida Cooperative Extension Service, Gulf Coast Research and Education Center, Bradenton, FL 34203
W. Atlee Burpee Company, 300 Park Avenue, Warminster, PA 18974

Some seed suppliers will also ship transplants, a practice that might be suitable for some if transplants are not locally available. Suppliers of transplant by mail are

Santa Barbara Heirloom Seedling Nursery, P.O. Box 4235, Santa Barbara, CA 93140
The Natural Gardening Company, 217 San Anselmo Avenue, San Anselmo, CA 94960
W. Atlee Burpee Company, 300 Park Avenue, Warminster, PA 18974

INTERNET SEED SOURCE

The 1997 Tomato Source Guide at Sherry's Greenhouse provides characteristic information for 301 tomato varieties. The tomato guide may be found at:

http://www.telport.com/~earth/GH301tombeef.html

CATALOG DESCRIPTIONS

Catalog descriptions normally give the variety, type of fruit (beefsteak, medium to large, small to medium, cherry, or paste), fruit color (red, yellow, orange), growth habit (determinate or indeterminate), days to fruiting [early (40–60 days) or full season (>65 days)], resistance to disease and other pests, adaptation to varying climatic conditions, and number or weight of seeds. A typical catalog description (from Totally Tomatoes) for one variety is

Better Boy Hybrid VFNASt—75 days.
Plump, juicy, deep red fruits, often over a pound. This is one of our most popular hybrids. Highly adaptable, produces full-season. Fine disease resistance. Indeterminate.

The average days for maturity from transplanting to harvest fall into one of three categories:

Category	Days
Early	50–65
Midseason	70–80
Late	85–95

Normally the large-fruited varieties fall into the "late" category, while most of the determinate varieties fall into the "early" category. There are very early determinate varieties now available that have 45–50 day maturities and can withstand low temperatures. A catalog description for one of these new varieties has been taken from Gurney's Seed and Nursery 1998 Spring Catalog:

SUB-ARCTIC PLENTY, *World's Earliest Tomato*
Canadian-bred variety produces reliable yields in only 6 weeks! It's a fine choice for northern growers and a boon to southern gardeners where a full crop can be harvested long before insects and fungal diseases become a problem. Determinate vines bear in concentration center clusters—firm, tasty tomatoes 2 inches in diameter. Approx. 75 seeds per pkt. 45 days.

The common disease resistance codes given for the tomato variety, usually found on the seed packet are

Code	Disease
V	Verticillium wilt
F	Fusarium wilt
FF	Fusarium, races 1 and 2
N	Nematodes
T	Tobacco mosaic virus
A	Alternaria stem canker
St	Stemphylium gray leaf spot

It is important that the seed source be reliable and that the seeds obtained be free from disease organisms. If the seeds have been treated with a pest chemical or other material, it should be so designated on the seed packet.

SAVING SEED FROM MATURE TOMATO FRUIT

It is possible to save seed from mature fruit as has been described by Erney (1998). However, if the tomato variety is a hybrid, saved seed will not come true, producing plants that will resemble the parent varieties. Although the tomato flower is self-pollinated, insects can carry pollen from one plant to another, which can result in seed that is the product of cross-pollination if various varieties of tomato are in the immediate area. Therefore, some seed will produce plants that are the result of the cross-pollination.

Procedure

1. Select overripe fruit (seeds in immature fruit are also immature and will not germinate).
2. Cut the fruit in half and squeeze the seeds into a jar. The gelatinous coating on the seed is what prevents seed germination and must be removed.
3. Add water to equal the amount of seeds and juice in the jar.
4. Let the jar set for 3 days at room temperature to ferment.
5. Remove the debris on the surface of the fermented liquid and then stir.
6. Let the jar stand because the viable seeds will fall to the bottom of the jar, and again remove the suspended material above the seeds on the bottom.
7. Pour the remaining solution slowly through a strainer to catch the viable seed.
8. Put the seeds on a plate to dry, occasionally turning seeds over and around to ensure even drying.
9. When seeds are dry, place them in an envelope and store in a cool dry place.

8 Pest Identification and Control

CONTENTS

RESISTANT CULTIVARS

Significant developments have been made to develop cultivars that are resistant or tolerant to the commonly occurring diseases, insects, and nematodes that can infest the tomato plant (Stevens and Rick, 1986). Peet (1996b) has identified these pests and the existence of resistant cultivars as shown in Table 8.1.

PLANT DISEASES

Considerable progress has been made in breeding disease resistance or tolerance to the more commonly occurring tomato plant viruses (Oshima, 1978) and diseases, such as the verticillium and fusarium wilts, tobacco mosaic virus, alternaria stem canker, stemphylium gray spot, septoria leaf spot, and bacterial speck (*Pseudomonas*) (Yang, 1978; Mohyuddin, 1985; Killebrew, 1996). Bacterial and fungal diseases affecting tomato in the tropics have been reviewed by Yang (1978). A compendium of tomato diseases has been written by Jones et al. (1991). The commonly occurring diseases affecting tomato, and their description and control are given in Table 8.2.

Those diseases that are seedborne can be controlled by heat treating the seed as follows:

Disease	Seed Treatment	
	Temperature [°F(°C)]	Time (min)
Bacterial canker, bacterial spot, bacterial speck	122 (50)	25
Anthracnose	132 (55.5)	30

TABLE 8.1
Tomato Cultivars with Insect and Disease Resistance

	Resistant Cultivars Exist	Cultivars
	Insects	
Aphid	No	
Cabbage looper	No	
Colorado potato beetle	No	
Fall armyworm	No	
Corn earworm	No	
Leaf miners	No	
Thrips	No	
Tomato pinworm	No	
	Diseases	
Alternaria stem canker	Yes	Common in hybrids
Anthracnose	No	
Bacterial canker	No	
Bacterial spot	No	
Bacterial speck	Yes	Some processing varieties
Bacterial wilt	Yes	Venus, Saturn, Kewado, Rosita
Early blight	Yes	Mountain Supreme
Fusarium wilt	Yes	Common for race 1 and race 2 gray mold open growth habit cultivars less prone to disease development
Gray leaf spot	Yes	Common in hybrids
Late blight	No	
Southern stem blight	No	
Tobacco mosaic	Yes	Common
Verticillium wilt	Yes	Common
Nematodes		
Root knot	Yes	Common

Source: Peet, M.M. 1996b. pp. 55–74. In: M.M. Peet (Ed.), *Sustainable Practices for Vegetable Production in the South*. Focus Publishing, R. Pullius Company, Newburyport, MA.

Whenever possible, growers should select those varieties that have disease resistance or tolerance if such disease organisms have been observed to be present in the past (see Table 7.1). Also, seed source must be considered to ensure that the seed obtained is disease free.

TABLE 8.2
Commonly Occurring Tomato Diseases

Disease	Description	Control
Anthracnose	Begins with circular, sunken spots on fruit; as spots enlarge, center becomes dark and fruit rots	Use approved fungicides
Bacterial canker	Wilting; rolling and browning of leaves; pith may discolor or disappear; fruit displays bird's-eye spots	Use hot-water-treated seed; avoid planting in affected fields for 3 years
Bacterial spot	Young lesions on fruit appear as dark, raised spots; older lesions blacken and appear sunken with brown centers; leaves brown and dry	Use hot-water-treated seed; use approved bactericides
Early blight	Dark brown spots on leaves; brown cankers on stems; girdling; dark, leathery, decayed areas at stem end of fruit	Use approved fungicides
Late blight	Dark, water-soaked spots on leaves; white fungus on undersides of leaves; withering of leaves; water-soaked spots on fruit turn brown; disease is favored by moist conditions	Use approved fungicides
Fusarium wilt	Yellowing and wilting of lower, older leaves; disease affects whole plant eventually	Use resistant varieties
Gray leaf spot	Symptoms appear first in seedlings; small brown to black spots on leaves, which enlarge and have shiny gray centers; centers may drop out to give shotgun appearance; oldest leaves affected first	Use resistant varieties; use approved fungicides
Leaf mold	Chlorotic spots on upper side of oldest leaves appear in humid weather; underside of leaf spot may have green mold; spots may merge until entire leaf is affected; disease advances to younger leaves	Use resistant varieties; stake and prune to provide air movement; use approved fungicides
Mosaic	Mottling (yellow and green) and roughening of leaves; dwarfing; reduced yields; russeting of fruit	Avoidance of contact by smokers; control of aphid carrier with insecticides; stylet oil
Verticillium wilt	Differs from fusarium wilt by appearance of disease on all branches at the same time; yellow areas on leaves become brown; midday wilting; leaves drop beginning at bottom	Use resistant varieties

Source: Maynard, D.H. and G.J. Hochmuth. 1997. *Knott's Handbook for Vegetable Growers.* 4th ed. John Wiley & Sons, New York.

Disease control requires constant plant observation and evaluation as new strains appear from the introduction of disease organisms from outside sources. For example, the outbreak of a new strain of the late blight pathogen (*Phytophthora infestans*), which has been introduced from Mexico, poses a serious problem to the tomato industry in Florida (Weingartner, 1997), requiring specific control measures by growers.

Gray mold (*Botrytis cinerea*) is a commonly occurring disease in warm and wet environments that primarily affects older leaf tissue. Its control is best done by keeping the foliage dry and by maintaining dry air movement within the plant canopy. Powell (1995) gives advice on how to control *Botrytis* environmentally as well as offering various fungicidal *Botrytis* strategies.

A soilborne disease that can be severe in warm wet weather conditions is *Rhizotonia,* which is best controlled by soil sterilization and by keeping soil from making contact with plant foliage.

A serious root disease is *Phytium aphanidermatum,* which occurs in warm wet conditions, frequently occurring in those hydroponic growing systems in which the nutrient solution is recirculated. When occurring, the disease can quickly kill plants, and its control requires dismantling of the entire hydroponic growing system for complete sterilization. In soilless medium systems, pine bark as an ingredient in the soilless mix offers control of this root disease.

Soilborne Diseases

Sterilization is required when tomato plants are grown in soil to control such diseases as bacterial wilt, southern blight, fusarium wilt, and fusarium crown rot. Field soil sterilization with methyl bromide has been a commonly used procedure; and for greenhouse soils, steam sterilization is used (Wittwer and Honma, 1969). Since use of methyl bromide is being phased out, alternative methods are being sought to replace it (Chellemi, 1997).

NEMATODES

Nematodes that affect tomato roots are root knot (*Meloidogyne* spp.), sting (*Belonoclaimus* spp.), and stubby root (*Trichodorus* spp.). Valdez (1978) lists those nematodes that attack tomatoes in the tropics. There are tomato plant varieties that are resistant to the root knot nematode (Stevens and Rick, 1986). Chemical nematode control is widely used; those chemicals registered for use in Florida are listed by Noling (1997). Nematode control under tropical conditions is given by Valdez (1978).

INSECTS

Insects that will attack the tomato plant are becoming increasingly difficult to control, particularly in the field and in those areas of the United States where tomato crops have been produced for a long period of time. The insect pressures in Florida, for example, have had a significant impact on the fresh market supply from that state, the major insect of concern being the silverleaf whitefly (*Bemisia argentifolli* Bellows & Perring) (Schuster, 1997). In addition, insects can carry diseases, such as various geminiviruses, which are just now being specifically identified (Polston and Anderson, 1997).

TABLE 8.3
Insects that Can Affect Tomato Plants

Insects	Description
Aphids	Small, green, pink, or black soft-body insects that rapidly reproduce to large populations; damage resulting from sucking plant sap, and indirectly from transmitting virus to crop plants
Colorado potato beetles	Oval beetles (3/8 in.) with ten yellow and ten black stripes that lay yellow eggs on undersides of leaves; brick-red, humpbacked larvae (1/2 in.) having black spots; beetles and larvae destructive leaf feeders
Corn earworms (tomato fruitworms)	Gray-brown moths (1 1/2 in.) with dark wing tips depositing eggs, especially on fresh corn silk; brown, green, or pink larvae (2 in.) feeding on silk, kernels, and foliage
Flea beetles	Small (1/6 in.) variable-colored, usually dark beetles, often present in large numbers in the early part of the growing season; feeding resulting in numerous small holes, giving a shotgun appearance; indirect damage resulting from diseases transmitted
Fruit flies	Small, dark-colored flies usually associated with overripe or decaying vegetables
Hornworms	Large (4–5 in.) moths that lay eggs that develop into large (3–4 in.) green fleshy worms with prominent white lines on sides and a distinct horn at the rear; voracious leaf feeders
Leaf miners	Tiny, black and yellow adults; yellowish-white maggotlike larvae tunneling within leaves and causing white or translucent, irregularly damaged areas
Pinworms	Tiny yellow, gray, or green, purple-spotted, brown-headed caterpillars that cause small fruit lesions, mostly near calyx; presence detected by large white blotches near folded leaves
Spider mites	Reddish, yellowish, or greenish, tiny, eight-legged spiders that suck plant sap from leaf undersides, causing distortion; fine webs possibly visible when mites are present in large numbers; mites not true insects
Stick bugs	Large, flattened, shield-shaped, bright green bugs; various-sized nymphs with reddish markings
Whiteflies	Small, whiteflies that move when disturbed

The insects that commonly affect tomato plants are given in Table 8.3.

Another insect that is becoming increasingly difficult to control is the thrip, a very small insect that can damage the tomato plant as well as carry virus diseases. Harris (1998) has described the best control measures to control thrips, measures such as screening ventilation openings and the use of insecticides.

Chemical insect control procedures, including chemicals and recommended application rates that would have wide application, have been given for Florida by Johnson (1997) and for Georgia by Guillebeau (1997). Papadopoulos et al. (1997), Snyder (1997b), and Gill and Sanderson (1998) discuss insect as well as disease control measures applicable to greenhouse tomato production.

Insect Control Using Predator Insects

The use of predator insects to control various insect pests has a fairly long history of investigation and development. The book edited by Hussey and Scopes (1985) and that by Malais and Ravensberg (1992) review the use of predators for biologically controlling insects in the greenhouse. The use of *Encarsia formosa* to control the caster whitefly in the greenhouse, for example, is becoming a commonly used control procedure (Smith, 1993; Ferguson, 1996; Stephens, 1997). Other insects are controllable using predator insects, as has been reviewed by Berlinger (1986) and Kueneman (1996).

Some of the plant-damaging insects and some of their predators are

Plant-Damaging Insects	Predators
Aphids	*Aphelinus abdominalis, Chrysoperia rufilabris*
Mealybugs	*Cryptolaemus montrouzieri*
Scales	*Aphytis melinus*
Spider mites	*Amblyseinus fallacis, Phytoseiulus perimilis*
Thrips	*Neoseiulus cucumeris*
Whiteflies	*Delphastus pusillus, Encarsia formosa*

The success in controlling insect populations using predators is based on careful monitoring and introduction of predators before the target insects become out of control. Temperature and humidity are important factors that can influence both the target insect and its predator, and therefore must be maintained at optimum levels. As with any pest problem, a combination of control factors becomes essential for success in keeping damaging insect populations from reaching detrimental levels (Ferguson, 1996).

Predator insects and instructions in their use are available from a number of suppliers.

Biological Control Sources

Koppert Biological Systems, 2856 South Main Street, Ann Arbor, MI 48103
IPM Laboratories, Inc., P.O. Box 300, Locke, NY 13092-0300
Troy Biosciences, 2620 North 37th Drive, Phoenix, AZ 85009
Envirepel-CAL CROP, USA, P.O. Box 4622, Escondido, CA 92046-2274

Biopesticides

A new line of pesticide chemicals is being developed containing naturally occurring fungus organisms that can invade the insect's body. One these products is BotaniGard™, a microinsecticide that can be used to control whiteflies (Stephens, 1997), thrips, and aphids. According to the manufacturer's (Mycotech Corporation, 529 East Front Street, Butte, MT 59702) literature, BotaniGard™ is "safe for workers and the environment, no preharvest interval required, exempt from all residue tolerances, IPM compatible, short reentry period, compatible with beneficial insects, and compatible with virtually all insecticides." Such a new chemical means of controlling insect pests in the tomato greenhouse will be welcome news to all growers. Such products are quite new and use experience by growers has not been well documented.

Insecticidal Soap

Insecticidal soap is another control chemical that will kill aphids, mealybugs, whiteflies, and mites. The plant needs to completely wetted with the insecticidal soap to be effective, and repeated applications may be needed to control an insect infestation.

Another form of an insecticidal soap is Organica® neem oil insecticidal soap (Organic, Inc., 111 Great Neck Road, Suite 320, Great Neck, NY 11021), a very new product, which is advertised to control a wide range of insects commonly occurring in the greenhouse. Neem-containing materials have had a long history for use in various developing countries around the world. Neem oil is extracted from the fruit of the neem (*Azadirachta indica*) tree. This product has yet to be tried by growers in sufficient numbers to determine its effectiveness in controlling insects found on the tomato in the greenhouse.

All these substances would be useful for cleaning the greenhouse between crops, washing the entire structure, floor, etc. with solutions of these various materials to kill any insects that may be residing in the greenhouse.

WEED CONTROL

Chemical soil weed control is the normal practice, methyl bromide (Gilbreath et al., 1997) being a control method when soils are sterilized with this agent. Pre- and post-planting weed control chemicals are frequently used depending on which weeds will or do occur. Peet (1996c) covers weed control procedures applicable in the southern United States. Common weeds and their chemical control under Florida conditions are described by Stall and Gilbreath (1997); and those under Georgia conditions, by Guillebeau (1997). Lange et al. (1986) have reviewed the various technical aspects of weed control including weed biology, field management, and chemical control (lists of chemicals need to be varied before use). Although herbicides can be effective in controlling weeds, cultural practices, such as crop rotation, tillage practices, and seedbed preparation, are equally important means of control. Therefore, a combined strategy of nonchemical and chemical techniques can effectively control most weed problems.

CHEMICAL PEST CONTROL

Most diseases, insects, nematodes, and weeds can be controlled chemically, although the use of chemicals for control can render fruit less desirable in the marketplace. Pest chemicals vary widely in their effectiveness, method and time of application, and requirement that they be applied only by those licensed to do so based on state or federal laws. Today, both federal and state laws require registration of most pest chemicals for regulation of their use. Therefore, the user of these chemicals needs to be aware of these regulations and follow label restrictions and instructions.

Chemicals that are approved for specific use are constantly changing as older products are removed from the market and new ones are introduced. A label can be changed by removing the use of a particular commonly known chemical for a particular crop or pest, although the chemical itself is still available for use on other crops or pests. Therefore, before selecting and using any pest chemical, its use must conform to label restrictions and requirements. Use and restriction information concerning pest chemicals can be found in various state agency publications, such as the *1997 Georgia Pest Control Handbook* (Guillebeau, 1997); 1997 Florida Tomato Institute Proceedings (Vavrina et al., 1997b); or the *Farm Chemical Handbook* (Anon., 1998c), the latter being a yearly publication that lists the label requirements on all pest chemicals. Growers need to have access to the latest versions of such publications for their growing region to ensure that the label is still applicable for its intended use.

Although most diseases produce characteristic visual symptoms, proper identification is essential by a skilled plant pathologist before a corrective chemical treatment is made. Many colleges of agriculture or their agricultural extension services within the land-grant university system in the United States and some soil and plant analysis laboratories (Anon., 1998b) offer pathological services and provide pest chemical recommendations. In addition, there are crop consultants in most of the major crop-producing areas of the United States who can field identify disease and insect pests and are usually familiar with current pest control regulations. In addition, some crop consultants are also licensed pest chemical applicators.

In all forms of pest control, the timing of applied control measures can determine the success or failure of treatment or treatments. Strategies for dealing with insects (Berlinger, 1986) and diseases (Watterson, 1986) need to be formulated and carefully followed. For example, it is important to know the life cycle of a plant-damaging insect to apply control measures that will prevent the development of that form of the insect that does the most harm to the crop. Each stage of insect development is different, and a particular control measure may be effective against one stage of development but not against another. As with disease control, professional assistance is important for correctly identifying the insect in question and then applying a control measure that will be effective.

INTEGRATED PEST MANAGEMENT

Integrated pest management (IPM) involves the use of a combination of procedures, cultural, and chemical and nonchemical, to control pests whether they be diseases (Watterson, 1986; Shipp et al., 1991; Peet, 1996b), insects (Berlinger, 1986; Shipp et al., 1991; Waterman, 1996; Peet, 1996a), or weeds (Lange et al., 1986; Peet, 1996c). An IPM program is normally for a specific crop, such as tomato (Rude, 1985; Anon., 1990; Ferguson, 1996); or for a system of growing, such as hydroponics (McEno, 1994). The application of IPM procedures is now the accepted method for pest control (Clarke et al., 1994).

Carefully maintaining the plant environment and the health of the plant itself is an important nonchemical means of disease and insect control. Air and rooting temperature, relative humidity, air flow, soil moisture, and plant nutrition, for example, are factors that contribute to the well-being of plants; these subjects are discussed in some detail in the book edited by Porter and Lawlor (1991). The silicon (Si) content of the plant has been found to be a factor in disease infestation as has been reported by Bélanger et al. (1995). A vigorously growing healthy plant is less likely to be disease prone. Stressed plants are more likely to be affected by the presence of disease and insects than a healthy plant is. Therefore, the focus of an IPM program needs to reach beyond chemical procedures.

PROPHYLACTIC PROCEDURES

As with any disease or insect infestation, prophylactic procedures are essential to prevent their introduction because most disease and insect problems are difficult to control after the fact. The environment, such as air and soil temperature, soil moisture, relative humidity, air movement within the plant canopy; the presence of host plants; and cultural practices will influence the initiation or control of diseases and insects. The use of insect traps and frequent monitoring of the tomato plant can warn of a developing insect or disease infestation so that control measures can be taken before the infestation reaches plant-damaging levels.

For the greenhouse grower, sanitation practices, and temperature and humidity control are important prevention procedures as well as screening to keep insects from being drawn into the greenhouse through the ventilation system. Controlling access to the greenhouse, sterilization of clothes and tools, etc. are equally important measures. Only disease- and insect-free plant material should be brought into the greenhouse. By combining chemical, nonchemical, and prophylactic procedures, a grower should be able to produce a tomato crop free from damaging pests.

References

Abbott, J.D., M.M. Peet, D.H. Willits, D.C. Sanders, and R.E. Gough. 1986. Effects of irrigation frequency and scheduling on fruit production and radial fruit cracking in greenhouse tomatoes in soil beds and in a soilless medium in bags. *Scientia Hortic.* 28:209–219.

Abeles, F.B., P.W. Morgan, and M.E. Saltveit. 1992. *Ethylene in Plant Biology.* Academic Press, San Diego, CA.

Adams, C.F. and M. Richardson. 1977. *Nutritive Value of Foods.* USDA-ARS Home and Garden Bulletin Number 72, U.S. Government Printing Office, Washington, D.C.

Adams, P. 1986. Mineral nutrition, pp. 281–334. In: J.G. Atherton and J. Rudich (Eds.), *The Tomato Crop: A Scientific Basis for Improvement.* Chapman & Hall, New York.

Adams, P. 1990. Effects of watering on the yield, quality and composition of tomatoes grown in bags of peat. *J. Hortic. Sci.* 65(6):667–674.

Adams, P. 1991. Effects of increasing the salinity of the nutrient solution with major nutrients or sodium chloride on the yield, quality, and composition of tomatoes grown in rockwool. *J. Hortic. Sci.* 66(2):207–210.

Adi Limited. 1982. Aeroponics in Israel. *HortScience* 17(2):137.

Adones, P. 1990. Effects of watering in the yield, quality and composition of tomatoes grown in bags of peat. *J. Hortic. Sci.* 65:667–674.

Ames, M. and W.S. Johnson. 1986. A review of factors affecting plant growth, Section V. In: *Proceedings 7th Annual Conference on Hydroponics: The Evolving Art, The Evolving Science.* Hydroponic Society of America, Concord, CA.

Ammerlaan, J.C.J. 1994. Environment-conscious production in glasshouse horticulture in the Netherlands. *Acta Hortic.* 36:67–76.

Anon. 1976. *United States Standards for Grades of Fresh Tomatoes.* United States Department of Agriculture, Agricultural Marketing Service, Washington, D.C.

Anon. 1978. *Nutritive Values of Fruits and Vegetables.* United Fresh Fruit and Vegetable Association, Alexandria, VA.

Anon. 1990. *Integrated Pest Management—Tomatoes.* 3rd ed. Publication 3274, Division of Agriculture and Natural Resources, University of California, Oakland, CA.

Anon. 1991. *Best Management Practices Begin with the Diagnostic Approach.* Potash and Phosphate Institute, Norcross, GA.

Anon. 1996. *Proceedings of the Greenhouse Tomato Seminar.* ASHS Press, American Society for Horticultural Science, Alexandria, VA.

Anon. 1997a. Water: What's in it and how to get it out. *Today's Chem.* 6(1):16–19.

Anon. 1997b. Dwarf gold treasurer: The longest keeper and the best tasting. *Org. Gardening* 44:37–38.

Anon. 1998a. *Tomato—Nutrient Deficiency Identification Guide.* Stoller Enterprises, Houston, TX.

Anon. 1998b. *Soil and Plant Analysis Laboratory Registry for the United States and Canada.* CRC Press, Boca Raton, FL.

Anon. 1998c. *Farm Chemical Handbook '98.* Meister Publishing, Willoughby, OH.

Anon. 1998d. Fruitful facts: News and notes of interest. *New Garden J.* 5(2):2.

Anon. 1998e. If you care about organic food...Act now. *Org. Gardening* 45(3):22–25.

Arnon, D.I. and P.R. Stout. 1939. The essentiality of certain elements in minute quantity for plants with special reference to copper. *Plant Physiol.* 14:371–375.

Asher, C.J. and D.G. Edwards. 1978. Critical external concentrations for nutrient deficiency and excess, pp. 13–28. In: A.R. Ferguson, R.L. Bialaski, and I.B. Ferguson (Eds.), *Proceedings 8th International Colloquium Plant Analysis and Fertilizer Problems.* Information Series No. 134, New Zealand Department of Scientific and Industrial Research, Wellington, New Zealand.

Asher, C.J. and J.F. Loneragan. 1963. Response of plants to phosphate concentrations in solution. I. Growth and phosphorus content. *Soil Sci.* 103:225–233.

Atherton, J.G. and G.P. Harris. 1986. Flowering, pp. 167–200. In: J.G. Atherton and J. Rudich (Eds.), *The Tomato Crop: A Scientific Basis for Improvement.* Chapman & Hall, New York.

Atherton, J.G. and J. Rudich (Eds.). 1986. *The Tomato Crop: A Scientific Basis for Improvement.* Chapman & Hall, New York.

Aung, L.H. 1978. Temperature regulation of growth and development of tomato during ontogeny, pp. 79–93. In: R. Cowell (Ed.), *1st International Symposium on Tropical Tomato.* Asian Vegetable Research and Development Center, Shanhua, Taiwan, Republic of China.

Baisden, G. 1994. Frankenfood: Bioengineered bonanza of the future, or your worst nightmare come true? *The Growing Edge* 5(4):34–37, 40–42.

Barber, S.A. 1995. *Soil Nutrient Bioavailability: A Mechanistic Approach.* 2nd ed. John Wiley & Sons, New York.

Barber, S.A. and D.R. Bouldin (Eds.). 1984. *Roots, Nutrients and Water Influx, and Plant Growth.* ASA Publication 136, American Society of Agronomy, Madison, WI.

Barker, A.V. and H.A. Mills. 1980. Ammonium and nitrate nutrition of horticultural crops, pp. 395–423. In: J. Janick (Ed.), *Horticultural Reviews.* AVI Publishing, Westport, CT.

Bartholomew, M. 1981. *Square Foot Gardening.* Rodale Press, Emmaus, PA.

Basiouny, F.M., K. Basiouny, and M. Maloney. 1994. Influence of water stress on abscisic acid and ethylene production in tomato under different PAR levels. *J. Hortic. Sci.* 69(3):535–541.

Bauerle, W.L. 1984. *Bag Culture Production of Greenhouse Tomatoes.* Ohio State University Special Circular 108, Ohio State University, Columbus, OH.

Bauerle, W.L. 1990. A window into the future in precision nutrient control, pp. 25–27. In: S. Korney (Ed.), *Proceedings of the 11th Annual Conference on Hydroponics.* Hydroponic Society of America, Concord, CA.

Bauerle, W.L., T.H. Short, E. Mora, S. Hoffman, and T. Nantais. 1988. Computerized individual nutrient fertilizer injector. The System. *HortScience* 23(5):910.

Bélanger, R.R., P.A. Bowen, D.L. Ehret, and J.G. Menzies. 1995. Soluble silicon: Its role in crop and disease management of greenhouse crops. *Plant Dis.* 79(4):329–335.

Benjamin, J. 1997. Paste tomato are plum tasty. *Org. Gardening* 44(4):28–32.

Bennett, W.F. (Ed.). 1993. *Nutrient Deficiencies and Toxicities in Crop Plants.* APS Press, American Phytopathological Society, St. Paul, MN.

Bergmann, W. 1992. *Nutritional Disorders of Plants: Developments, Visual and Analytical Diagnosis.* Gustav Fischer Verlag, Jenna, Germany.

Berlinger, M.J. 1986. Pests, pp. 391–441. In: J.G. Atherton and J. Rudich (Eds.), *The Tomato Crop: A Scientific Basis for Improvement.* Chapman & Hall, New York.

Berry, W.L. 1989. Nutrient control and maintenance in solution culture, pp. 1–6. In: S. Korney (Ed.), *Proceedings of the 10th Annual Conference on Hydroponics.* Hydroponic Society of America, San Ramon, CA.

Berry, W.L. and A. Wallace. 1981. Toxicity: The concept and relationship to the dose response curve. *J. Plant Nutr.* 3:13–19.

Beverly, R.B. 1994. Stem sap testing as a real-time guide to tomato seedling nitrogen and potassium fertilization. *Commun. Soil Sci. Plant Anal.* 25:1045–1056.

Bij, J. 1990. Growing commercial vegetables in rockwool, pp. 18–24. In: S. Korney (Ed.), *Proceedings of the 11th Annual Conference on Hydroponics.* Hydroponic Society of America, Concord, CA.

Bloom, A. 1987. Nutrient requirement changes during plant development, pp. 104–112. In: *Proceedings of the 8th Annual Conference on Hydroponics*—Effective Growing Techniques. Hydroponic Society of America, Concord, CA.

Boodley, J.W. and R. Sheldrake, Jr. 1972. *Cornell Peat-Lite Mixes for Commercial Plant Growing.* Information Bulletin 43, Cornell University, Ithaca, NY.

Brentlinger, D. 1992. Tomatoes in perlite: A simplified hydroponic system. *Am. Veg. Grower* 40:51–52.

Brentlinger, D. 1997. Commercial hydroponics in North America, pp. 67–73. In: R. Wignarajah (Ed.), *Proceedings 18th Annual Conference on Hydroponics.* Hydroponic Society of America, San Ramon, CA.

Brooks, W.M. 1969. *Growing Greenhouse Tomatoes.* SB-19. Cooperative Extension Service, Ohio State University, Columbus, OH.

Brown, P.H., R.M. Welsh, and E.E. Cary. 1987. Nickel: A micronutrient essential for higher plants. *Plant Physiol.* 85:801–803.

Bruce, R.R, J.E. Pallas, Jr., L.A. Harper, and J.B. Jones, Jr. 1980. Water and nutrient element regulation prescription in nonsoil media for greenhouse crop production. *Commun. Soil Sci. Plant Anal.* 11(7):677–698.

Bugbee, B. 1995. Nutrient management in recirculating hydroponic culture, pp. 15–30. In: M. Bates (Ed.), *Proceedings of the 16th Annual Conference on Hydroponics.* Hydroponic Society of America, San Ramon, CA.

Bunt, A.C. 1988. *Media and Mixes for Container-Grown Plants.* 2nd ed. Unwin Hyman, London, England.

Burnham, M., F. Killebrew, and P. Harris. 1996. *Garden Tabloid.* Cooperative Extension Service, Mississippi State University, Mississippi State, MS.

Buyanovsky, G., J. Gale, and N. Degani. 1981. Ultra-violet radiation for the inactivation of microorganisms in hydroponics. *Plant Soil* 60:131–136.

Cantliffe, D.J. 1997. Impression of west Mexican fresh market vegetable agriculture, Pre and post NAFTA (1982–1997), pp. 5–6. In: C.S. Vavrina, P.R. Gilreath, and J.W. Noling (Eds.), *1977 Florida Tomato Institute Proceedings.* Citrus and Vegetable Magazine, University of Florida, Gainesville, FL.

Carrier, André. 1997. Status of the Quebec greenhouse industry, pp. 181–186. In: K. Wignarajah (Ed.), *Proceedings 18th Annual Conference on Hydroponics.* Hydroponic Society of America, San Ramon, CA.

Carson, E.W. (Ed.). 1974. *The Plant Root and Its Environment.* University Press of Virginia, Charlottesville, VA.

Carter, M.R. (Ed.). 1993. *Soil Sampling and Methods of Analysis.* Lewis Publishers, Boca Raton, FL.

Cavagnaro, D. 1996. Grow tomatoes by the ton. *Org. Gardening* 43(1):46–51.

Cebenko, J.J. 1997. Tomatoes that store long and taste great! *Org. Gardening* 44(7):36–38.

Chellemi, D. 1997. Alternatives to methyl bromide for soilborne disease management, pp. 36–37. In: C.S. Vavrina, P.R. Gilreath, and J.W. Noling (Eds.), *1997 Florida Tomato Institute Proceedings.* Citrus and Vegetable Magazine, University of Florida, Gainesville, FL.

Clark, G.A. and A.G. Smajstrla. 1996a. Design considerations for vegetable crop drip irrigation systems. *HortTechnology* 6(3):155–159.

Clark, G.A. and A.G. Smajstrla. 1996b. Injecting chemical into drip irrigation systems. *HortTechnology* 6(3):160–165.

Clarke, N.D., J.L. Shipp, W.R. Jarvis, A.P. Papadopoulos, and T.J. Jewett. 1994. Integrated management of greenhouse crops—A conceptual and potentially practical model. *HortScience* 29:846–849.

Cockshull, K.E. and L.C. Ho. 1995. Regulation of tomato fruit size by plant density and truss thinning. *J. Hortic. Sci.* 70:395–407.

Cockshull, K.E., C.J. Graves, and C.R.J. Cave. 1992. The influence of shading on yield of glasshouse tomatoes. *J. Hortic. Sci.* 67:11–24.

Coltman, R.R. 1987. Yield and sap nitrate responses of fresh market field tomatoes to simulated fertigation with nitrogen. *J. Plant Nutr.* 10:1699–1704.

Coltman, R.R. 1988. Yields of greenhouse tomatoes managed to maintain specific petiole sap nitrate levels. *HortScience* 23:148–151.

Cooper, A. 1996. *The ABC of NFT, Nutrient Film Technique*. Casper Publications, Narrabeen, Australia.

Cowell, R. 1979. *Proceedings of the 1st International Symposium on Tropical Tomato.* AVRDC Publication 78-59, Asian Vegetable Research and Development Center, Shanhua, Taiwan, Republic of China.

Curry, H. 1997. Colorado greenhouses: State-of-the-art tomato production. *Greenhouse Manage. Prod.* 17:30–34.

Davies, J.N. and G.E. Hobson. 1981. The constituents of tomato fruit—the influence of environment, nutrition, and genotype. *CRC Crit. Rev. Food Sci. Nutri.* 15:205–280.

Davis, N. 1996. Lighting for plant growth: An overview, pp. 59–69. In: K. Wignarajah (Ed.), *Proceedings of the 17th Annual Conference on Hydroponics.* Hydroponic Society of America, San Ramon, CA.

Day, D. 1991. *Growing in Perlite*. Grower Digest 12, Grower Books, London, England.

De Koning, A.N.M. 1989. Development and growth of a commercially grown tomato crop. *Acta Hortic.* 260:267–273.

DeKock, P.C., R.H.E. Inkson, and A. Hall. 1982. Blossom-end rot of tomato as influenced by truss size. *J. Plant Nutr.* 5:57–62.

De Kreij, C., J. Janse, B.J. van Goor, and J.D.J. van Doesburg. 1992. The incidence of calcium oxalate crystals in fruit walls of tomato (*Lycopersicon esculentum* Mill) as affected by humidity, phosphate, and calcium supply. *J. Hortic. Sci.* 67:45–50.

Dieleman, J.A. and E. Heuvelink. 1992. Factors affecting the number of leaves preceding the first inflorescence on the tomato. *J. Hortic. Sci.* 67:1–10.

Downing, R. 1997. *1997–1998 Compost/Organics Reference Book*. Downing & Associates, Mentor, OH.

Eastwood, T. 1947. *Soilless Growth of Plants*. Reinhold Publishing, New York.

Edwards, R. 1994. Computer control systems wired to grow. *Growing Edge* 5(3):34–38.

Ehret, D.L. and L.C. Ho. 1986. The effects of salinity on dry matter partitioning and fruit growth in tomatoes grown in nutrient film culture. *J. Hortic. Sci.* 61(3):361–367.

Epstein, E. 1965. Mineral nutrition, pp. 438–466. In: J. Bonner and J.E. Varner (Eds.), *Plant Biochemistry*. Academic Press, Orlando, FL.

Epstein, E. 1972. *Mineral Nutrient of Plants: Principles and Perspectives.* John Wiley & Sons, New York.

Epstein, E. 1994. The anomaly of silicon in plant biology. *Proc. Natl. Acad. Sci. (USA)* 91:11–17.

Erney, D. 1998. Save seed from your favorite tomato varieties. *Org. Gardening* 45(6):68.

Eskew, D.L., R.M. Welsh, and W.A. Norvell. 1984. Nickel in higher plants: Further evidence for an essential role. *Plant Physiol.* 76:691–693.

Evans, R.D. 1995. Control of microorganisms in flowing nutrient solutions, pp. 31–43. In: M. Bates (Ed.), *Proceedings 16th Annual Conference on Hydroponics.* Hydroponic Society of America, San Ramon, CA.

Farmhand, D.S., R.F. Hasek, and J.L. Paul. 1985. *Water Quality.* Leaflet 2995, Division of Agriculture Science, University of California, Davis, CA.

Ferguson, G. 1996. Management of greenhouse tomato pests: An integrated approach, pp. 26–30. In: *Greenhouse Tomato Seminar.* ASHS Press, American Society for Horticultural Science, Alexandria, VA.

Fischer, D.F., G.A. Giacomelli, and H.W. Janes. 1990. A system of intensive tomato production using ebb-flood benches. *Prof. Hortic.* 4(3):99–106.

Fortnum, B.A., D.R. Decoteau, and M.J. Kasperbauer. 1997. Colored mulches affect yield of fresh-market tomato infected with *Meloidogyne incognita. J. Nematol.* 29(4): 1–15.

Gallagher, P.A. 1972. Potassium nutrition of tomatoes, pp. 13–18. In: *Proceedings Provincial Glasshouse Conference*, Dublin, Ireland.

Geisenberg, C. and K. Stewart. 1986. Field crop management, pp. 511–557. In: J.G. Atherton and J. Rudich (Eds.), *The Tomato Crop: A Scientific Basis for Improvement.* Chapman & Hall, New York.

Geraldson, C.M. 1963. Quantity and balance of nutrients required for best yields and quality tomatoes. *Proc. Fla. State Hortic. Soc.* 76:153–158.

Geraldson, C.M. 1982. Tomato and the associated composition of the hydroponic or soil solution. *J. Plant Nutr.* 5(8):1091–1098.

Gerber, J.M. 1985. Plant growth and nutrient formulas, pp. 58–69. In: A.J. Savage (Ed.), *Hydroponics Worldwide: State of the Art in Soilless Crop Production.* International Center for Special Studies, Honolulu, HI.

Gerhart, K.A. and R.C. Gerhart. 1992. Commercial vegetable production in a perlite system, pp. 35–38. In: D. Schact (Ed.), *Proceedings of the 13th Annual Conference on Hydroponics.* Hydroponic Society of America, San Ramon, CA,

Giacomelli, G.A. 1996a. Designing the greenhouse to fit the needs of the plant, pp. 7–12. In: *Proceedings of the Greenhouse Tomato Seminar.* ASHS Press, American Society for Horticultural Science, Alexandria, VA.

Giacomelli, G.A. 1996b. Cooling and heating the greenhouse for control of plant environment, pp. 16–20. In: *Proceedings of the Greenhouse Tomato Seminar.* ASHS Press, American Society for Horticultural Science, Alexandria, VA.

Giacomelli, G.A., K.C. Ting, and D.R. Mears. 1993. *Design of a Single Truss Tomato Production System (STTPS).* Symposium on New Cultivation Systems, Cagliari, Italy.

Giacomelli, G.A. and K.C. Ting. 1994. *Greenhouse Systems: Automation, Culture and Environment.* Northeast Regional Agricultural Engineering Service Publication NRAES-72, NRAES, Cornell University, Ithaca, NY.

Gieling, Th.H. 1985. Electronics, sensors, and software for microelectronics in the greenhouses, pp. 36–41. In: A.J. Savage (Ed.), *Hydroponics Worldwide: State of the Art in Soilless Crop Production.* International Center for Special Studies, Honolulu, HI.

Gilbreath, J.P., J.P. Jones, J.W. Noling, and P.R. Gilbreath. 1997. Alternatives to methyl bromide for management of weeds, pp. 34–35. In: C.S. Vavrina, P.R. Gilreath, and J.W. Noling (Eds.), *1997 Florida Tomato Institute Proceedings.* Citrus and Vegetable Magazine, University of Florida, Gainesville, FL.

Gill, S. and J. Sanderson. 1998. *Ball Guide to Identification of Greenhouse Pests and Beneficials.* Ball Publishing Books, Batavia, IL.

Giniger, M.S., R.J. McAvoy, G.A. Giacomelli, and H.W. Janes. 1988. Computer simulation of a single truss tomato cropping system. *Trans. ASAE* 31(4):1176–1179.

Glass, A.D.M. 1989. *Plant Nutrition: An Introduction to Current Concepts.* Jones & Bartlett Publishers, Boston, MA.

Goldberry, K.L. 1985. Greenhouse structures and systems, pp. 12–20. In: A.J. Savage (Ed.), *Hydroponics Worldwide: State of the Art in Soilless Crop Production.* International Center for Special Studies, Honolulu, HI.

Gough, C. and G.E. Hobson. 1990. A comparison of the productivity, quality, self-life characteristics, and consumer reaction to the crop from cherry tomato plants grown at different levels of salinity. *J. Hortic. Sci.* 65(4):431–439.

Grierson, D. and A.A. Kader. 1986. Fruit ripening and quality, pp. 241–280. In: J.G. Atherton and J. Rudich (Eds.), *The Tomato Crop: A Scientific Basis for Improvement.* Chapman & Hall, New York.

Guillebeau, Paul (Ed.). 1997. *1997 Georgia Pest Control Handbook.* Special Bulletin 28, Georgia Cooperative Extension Service, Athens, GA.

Gunstone, G.W. 1994. Biological systems for glasshouse horticulture: Koppert. *Growing Edge* 5(4):43–47, 50–51.

Halbrooks, M.C. and G.E. Wilcox. 1980. Tomato plant development and elemental accumulation. *J. Am. Soc. Hortic. Sci.* 105:826–828.

Halliday, D.J. and M.E. Trenkel (Eds.). 1992. *IFA World Fertilizer Use Manual*, pp. 289–290, 331–337. International Fertilizer Industry Association, Paris, France.

Hanan, J.J. 1998. *Greenhouses: Advanced Technology for Protected Horticulture.* CRC Press, Boca Raton, FL.

Harper, L.A., J.E. Pallas, Jr., R.R. Bruce, and J.B. Jones, Jr. 1979. Greenhouse microclimate for tomatoes in the southeast. *J. Am. Soc. Hortic. Sci.* 104(5):659–663.

Harris, M. 1998. Tips for fine-tuning your trips control program. *GMPro* 18(1):56–57.

Harris, P., J.H. Jarratt, F. Killebrew, J. Byrd, Jr., and R.G. Snyder. 1996. *Organic Vegetable IPM Guide.* Cooperative Extension Publication 2036, Mississippi State University, MS.

Hartman, P.L., H.A. Mills, and J.B. Jones, Jr. 1986. The influence of nitrate:ammonium ratios on growth, fruit development, and element concentration in 'Floradel' tomato plants. *J. Am. Soc. Hortic. Sci.* 111(4):487–490.

Hartz, T.K. 1996. Water management in drip-irrigated vegetable production. *HortTechnology* 63(3):165–167.

Hartz, T.K. and G.J. Hochmuth. 1996. Fertility management of drip-irrigated vegetables. *HortTechnology* 6(3):186–172.

Heiser, C.B., Jr. 1969. *Nightshades, the Paradoxical Plants.* W.H. Freeman, San Francisco, CA.

Hendrickson, R. 1977. *The Great American Tomato Book: The One Complete Guide to Growing and Using Tomatoes Everywhere.* Doubleday, Garden City, NY.

Hessayon, D.G. 1997. *The New Vegetable and Herd Expert,* pp. 98–104. Expert Books, Transworld Publishers, London, England.

Heuvelink, E. and N. Bertin. 1994. Dry matter partitioning in a tomato crop: Comparison of two simulation models. *J. Hortic. Sci.* 69:885–903.

Ho, L.C. and J.D. Hewitt. 1986. Fruit development, pp. 201–239. In: J.G. Atherton and J. Rudich (Eds.), *The Tomato Crop: A Scientific Basis for Improvement.* Chapman & Hall, New York.

Hoagland, D.R. and D.I. Arnon. 1950. *The Water Culture Method for Growing Plants without Soil.* Circular 347, California Agricultural Experiment Station, University of California, Berkeley, CA.

Hochmuth, G.J. 1991. Production of greenhouse tomatoes in Florida. In: C.J. Hochmuth (Ed.), *Florida Greenhouse Vegetable Production Handbook*. Volume 3, University of Florida, Gainesville, FL.

Hochmuth, G.J. 1996a. Tomato fertilizer management, pp. vii–xv. In: C.S. Vavrina (Ed.), *1996 Proceedings Florida Tomato Institute*. PRO 108, University of Florida, Gainesville, FL.

Hochmuth, G.J. 1996b. Greenhouse tomato nutrition and fertilization for southern latitudes, pp. 37–39. In: *Greenhouse Tomato Seminar*. ASHS Press, American Society for Horticultural Science, Alexandria, VA.

Hochmuth, G.J. 1997. Tomato fertilizer management, pp. 68–69. In: C.S. Vavrina, P.R. Gilreath, and J.W. Noling (Eds.), *1977 Florida Tomato Institute Proceedings*. Citrus and Vegetable Magazine, University of Florida, Gainesville, FL.

Hochmuth, G.J. and B. Hochmuth. 1996. Challenges for growing tomatoes in warm climates, pp. 34–36. In: *Greenhouse Tomato Seminar*. ASHS Press, American Society for Horticultural Science, Alexandria, VA.

Hochmuth, R.C., L.L. Leon, and G.J. Hochmuth. 1997. Cluster tomatoes, pp. 16–17. In: C.S. Vavrina, P.R. Gilreath, and J.W. Noling (Eds.), *1977 Florida Tomato Institute Proceedings*. Citrus and Vegetable Magazine, University of Florida, Gainesville, FL.

Hoagland, D.R. and D.I. Arnon. 1950. *The Water-Culture Method for Growing Plants without Soil*. Circular 347. Agricultural Experiment Station, University of California, Berkeley, CA.

Hussey, N.W. and N. Scopes (Eds.). 1985. *Biological Pest Control: In Greenhouse Experience*. Blandford Press, Poole, Dorset, United Kingdom.

Ingratta, F.J., T.J. Blom, and W.A. Straver. 1985. Canada: Current research and developments, pp. 95–102. In: A.J. Savage (Ed.), *Hydroponics Worldwide: State of the Art in Soilless Crop Production*. International Center for Special Studies, Honolulu, HI.

James, Henry. 1996. *The Farmer's Guide to the Internet*. TVA Rural Studies, University of Kentucky, Lexington, KY.

Janes, H.W. 1994. Tomato production in protected cultivation. *Encycl. Agric. Sci.* 4:337–349.

Jensen, M.H. 1997. Hydroponics. *HortScience* 32(6):1018–1021.

Jensen, M.H. and A.J. Malter. 1995. *Protected Agriculture: A Global Review*. World Bank Technical Paper No. 253, World Bank, Washington, D.C.

Jesiolowski, J. 1996. Get bugs to boost your yields. *Org. Gardening* 43(5):28–35.

Johnson, F. 1997. Chemical insect control in tomatoes, pp. 73–83. In: C.S. Vavrina, P.R. Gilreath, and J.W. Noling (Eds.), *1997 Florida Tomato Institute Proceedings*. Citrus and Vegetable Magazine, University of Florida, Gainesville, FL.

Johnson, H., Jr., C.J. Hochmuth, and D.N. Maynard. 1985. *Soilless Culture of Greenhouse Vegetables*. IFAS Bulletin 218, C.M. Hinton Publications Distribution Center. Cooperative Extension Service, University of Florida, Gainesville, FL.

Jones, J.B., Jr. 1980. Construct your own automatic growing machine. *Pop. Sci.* 216(3):87.

Jones, J.B., Jr. 1993a. *Nutrient Element Deficiencies in Tomato*. (Video). St. Lucie Press, Boca Raton, FL.

Jones, J.B., Jr. 1993b. *Plant Nutrition Basics* (Video). St. Lucie Press, Boca Raton, FL.

Jones, J.B., Jr. 1993c. *Plant Analysis* (Video). St. Lucie Press, Boca Raton, FL.

Jones, J.B., Jr. 1993d. *Tissue Testing* (Video). St. Lucie Press, Boca Raton, FL.

Jones, J.B., Jr. 1997a. *Plant Nutrition Manual*. CRC Press, Boca Raton, FL.

Jones, J.B., Jr. 1997b. *Hydroponics: A Practical Guide for the Soilless Grower.* St. Lucie Press, Boca Raton, FL.

Jones, J.B., Jr. 1997c. Advantages gained by controlling root growth on a newly-developed hydroponic growing system, pp. 125–136. In: R. Wignarajah (Ed.), *Proceedings 18th Annual Conference on Hydroponics*, Hydroponic Society of America, San Ramon, CA.

Jones, J.B., Jr. 1998. Phosphorus toxicity in plants: When and how does it occur? *Commun. Soil Sci. Plant Anal.* 29(11–12): in press.

Jones, J.B., J.P. Jones, R.E. Stall, and T.A. Zitter (Eds.). 1991. *Compendium of Tomato Diseases.* APS Press, American Phytopathological Society, St. Paul, MN.

Jones, Scott. 1996. The hydroponic grower on-line, pp. 51–57. In: R. Wignarajah (Ed.), *Proceedings 17th Annual Conference on Hydroponics.* Hydroponic Society of America, San Ramon, CA.

Kabata-Pendias, A. and H. Pendias. 1994. *Trace Elements in Soils and Plants.* 2nd ed. CRC Press, Boca Raton, FL.

Kader, A.A. (Ed.). 1992. *Postharvest Technology of Horticultural Crops of California.* California Division of Agriculture and Natural Resources Publication 3311, Sacramento, CA.

Kalra, Y.P. (Ed.). 1997. *Handbook on Reference Methods for Plant Analysis.* CRC Press, Boca Raton, FL.

Keller, J. and R.D. Bliesner. 1990. *Sprinkle and Trickle Irrigation.* Van Nostrand Reinhold, New York.

Killebrew, F. 1996. Greenhouse tomato disease identification and management, pp. 21–25. In: *Greenhouse Tomato Seminar.* ASHS Press, American Society for Horticultural Science, Alexandria, VA.

Kinet, J.M. 1977. Effects of light conditions on the development of the inflorescence in tomato. *Sci. Hortic.* 6:15–26.

Knecht, G.N. and J.W. O'Leary. 1974. Increased tomato fruit development by CO_2-enrichment. *J. Am. Soc. Hortic. Sci.* 99:214–216.

Koske, T.J., J.E. Pallas, and J.B. Jones, Jr. 1980. Influence of ground bed heating and cultivar on tomato fruit cracking. *HortScience* 15:760–762.

Kueneman, T.C. 1996. Bumble bee pollination and biological control, pp. 107–111. In: K. Wignarajah (Ed.), *Proceedings of the 17th Annual Conference on Hydroponics.* Hydroponic Society of America, San Ramon, CA.

Kuo, C.G., B.W. Chen, M.H. Chou, C.L. Tsai, and T.S. Tsay. 1978. Tomato fruit-set at high temperature, pp. 94–109. In: R. Cowell (Ed.), *1st International Symposium on Tropical Tomato.* Asian Vegetable Research and Development Center, Shanhua, Taiwan, Republic of China.

Lamont, Jr., W.J. 1996. What are the components of a plasticulture vegetable system? *HortTechnology* 6(3):150–154.

Lange, A.H., B.B. Fischer, and F.M. Ashton. 1986. Weed control, pp. 485–510. In: J.G. Atherton and J. Rudich (Eds.), *The Tomato Crop: A Scientific Basis for Improvement.* Chapman & Hall, New York.

Leskovar, D.I. and D.J. Cantliffe. 1990. Does the initial condition of the transplants affect tomato growth and development? *Proc. Fla. State Hortic. Soc.* 103:148–153.

Lindsay, W.L. 1979. *Chemical Equilibria in Soils.* John Wiley & Sons, New York.

Logendra, L.S. and H.W. Janes. 1997. Hydroponic tomato production: Growing media requirements, pp. 119–123. In: K. Wignarajah (Ed.), *Proceedings 18th Annual Conference on Hydroponics.* Hydroponic Society of America, San Ramon, CA.

Lorenz, O.A. and D.N. Maynard. 1988. *Knott's Handbook for Vegetable Growers.* 3rd ed., John Wiley & Sons, New York.

Magalhaes, J.R. and G.E. Wilcox. 1984. Ammonium toxicity development in tomato plants relative to nitrogen form and light intensity. *J. Plant Nutr.* 7(10):1477–1496.

Magoon, C.E. 1978. *Supply Guide: Average Monthly Availability of Fresh Fruits and Vegetables.* United Fresh Fruit and Vegetable Association, Alexandria, VA.

Malais, M. and W.J. Ravensberg. 1992. *Knowing and Recognizing the Biology of Glasshouse Pests and Their Natural Enemies.* Koppert Biological Systems. Koppert B.V. Berkel en Rodenrijs, The Netherlands.

Manrique, L.A. 1993. Greenhouse crops: A review. *J. Plant Nutr.* 16(12):2411–2477.

Marschner, H. 1986. *Mineral Nutrition of Higher Plants.* Academic Press, New York.

Mattern, V. 1996. Strike Tomato Gold. *Org. Gardening* 43(3):31–35.

Maynard, D.H. 1997. Tomato varieties for Florida, pp. 66–67. In: C.S. Vavrina, P.R. Gilreath, and J.W. Noling (Eds.), *1997 Florida Tomato Institute Proceedings.* Citrus and Vegetable Magazine, University of Florida, Gainesville, FL.

Maynard, D.H. and G.J. Hochmuth. 1997. *Knott's Handbook for Vegetable Growers.* 4th ed. John Wiley & Sons, New York.

McAvoy, R.J. and H.W. Janes. 1984. The use of high pressure sodium lights in greenhouse tomato crop production. *Acta Hortic.* 148:877–884.

McAvoy, R.J. and H.W. Janes. 1988. Alternative production strategies for greenhouse tomatoes using supplemental lighting. *Sci. Hortic.* 35:161–166.

McAvoy, R.J. and H.W. Janes. 1989. Tomato plant photosynthesis activity as related to canopy age and tomato development. *J. Am. Soc. Hortic. Sci.* 114(3):478–482.

McAvoy, R.J., H.W. Janes, and G.A. Giacomelli. 1989a. Development of a plant factory model. I. The organization and operational model. II. A plant growth model: The single truss tomato crop. *Acta Hortic.* 248:85–94.

McAvoy, R.J., H.W. Janes, G.A. Giacomelli, and M.S. Giniger. 1989b. Validation of a computer model for a single-truss tomato cropping system. *J. Am. Soc. Hortic. Sci.* 114(5):746–750.

McEno, J. 1994. Hydroponic IPM, pp. 61–66. In: D. Parker (Ed.), *The Best of the Growing Edge.* New Moon Publishing, Corvallis, OR.

Melnick, R. 1998. Organizing organics. *Ag Consultant* 54:8.

Mengel, K. and E.A. Kirkby. 1987. *Principles of Plant Nutrition,* 4th ed. International Potash Institute, Bern, Switzerland.

Meyer, S. 1998. Starting tomatoes from seed. *Org. Gardening* 45(2):46–49.

Mills, H.A. and J. B. Jones, Jr. 1996. *Plant Nutrition Manual II.* Micro-Macro Publishing, Athens, GA.

Mirza, M. 1994. Managing the production of greenhouse vegetables for above average yields. In: *Texas Greenhouse Council Meeting,* July 1994. Fort Worth, TX.

Mirza, M. and M. Younus. 1997. An overview of the greenhouse vegetable industry in Alberta, Canada, pp. 187–189. In: K. Wignarajah (Ed.), *Proceedings 18th Annual Conference on Hydroponics.* Hydroponic Society of America, San Ramon, CA.

Mohyuddin, M. 1985. Crop cultivars and disease control, pp. 42–50. In: A.J. Savage (Ed.), *Hydroponics Worldwide: State of the Art in Soilless Crop Production.* International Center for Special Studies, Honolulu, HI.

Molyneux, C.J. 1988. *A Practical Guide to NFT.* T. Snap & Co., Preston, Lancashire, England.

Morard, P. and J. Kerhoas. 1984. Tomato and cucumber, pp. 677–687. In: P. Martin-Prével, J. Gagnard, and P. Gautier (Eds.), *Plant Analysis as a Guide to the Nutrient Requirements of Temperate and Tropical Crops.* Lavoisier Publishing, New York.

Morgan, L. 1997. Solutions for that home-grown flavor. *GrowingEdge* 8(4):24–31.

Mortvedt, J.J. 1991. *Micronutrients in Agriculture.* 2nd ed. SSSA Book Series No. 4, Soil Science Society of America, Madison, WI.

Moretti, C.L., A.S. Sargent, D.J. Huber, and R. Puschmann. 1997. Internal bruising affects chemical and physical composition of tomato fruits. *HortScience* 32(3):522.

Mozafar, A. 1993. Nitrogen fertilizers and the amount of vitamins in plants: A review. *J. Plant Nutr.* 16(12):2479–2506.

Mpelkas, C.C. 1989. Electric energy: A key environmental factor in horticultural technology, pp. 53–77. K.L. McFate (Ed.), *Electrical Energy in Agriculture*. Elsevier, Amsterdam, The Netherlands.

Muckle, M.E. 1990. *Hydroponic Nutrients—Easy Ways to Make Your Own*. Revised ed. Growers Press, Princeton, B.C., Canada.

Naegely, S.K. 1997. Greenhouse vegetables: Business is booming. *Greenhouse Grower* 15:14–18.

Nakayama, F.S. and D.A. Bucks. 1986. *Trickle Irrigation for Crop Production*. Elsevier Science Publishers, Amsterdam, The Netherlands.

Noling, J.W. 1997. Nematicides registered for use on Florida tomato, pp. 78, 92. In: C.S. Vavrina, P.R. Gilreath, and J.W. Noling (Eds.), *1997 Florida Tomato Institute Proceedings*. Citrus and Vegetable Magazine, University of Florida, Gainesville, FL.

Orzolek, M.D. 1996. Stand establishment in plasticulture systems. *HortTechnology* 6(3):181–185.

Oshima, N. 1978. Tomato viruses, pp. 124–131. In: R. Cowell (Ed.), *1st International Symposium on Tropical Tomato*. Asian Vegetable Research and Development Center, Shanhua, Taiwan, Republic of China.

Pais, I. and J.B. Jones, Jr. 1997. *The Handbook on Trace Elements*. St. Lucie Press, Boca Raton, FL.

Papadakis, G., A. Frangoudakis, and S. Kyritis. 1994. Experimental investigation and model of heat and mass transfer between a tomato crop and the greenhouse environment. *J. Agric. Eng. Res.* 57:217–227.

Papadopoulos, A.P. 1991. *Growing Greenhouse Tomatoes in Soil and in Soilless Media*. Agriculture Canada Publication 186/E, Communications Branch, Agriculture Canada, Ottawa, Canada.

Papadopoulos, A.P. and S. Pararajasingham. 1996. The influence of plant spacing on light interception and use in greenhouse tomato (*Lycopersicon esculentum* Mill.): A review. *Scienta Hortic.* 69:1–29.

Papadopoulos, A.P., S. Pararajasingham, J.L. Shipp, W.R. Jarvis, and T.J. Jewett. 1997. Integrated management of greenhouse vegetable crops. *Hortic. Rev.* 21:1–38.

Parker, D. (Ed.). 1994. *The Best of the Growing Edge*. New Moon Publishing, Corvallis, OR.

Parnes, R. 1990. *Fertile Soil, A Grower's Guide to Organic and Inorganic Fertilizers*. AgAccess, Davis, CA.

Peet, M.M. 1992. Fruit cracking in tomato. *HortTechnology* 2:216–223 [Errata. *HortTechnol.* 2(3):432].

Peet, M.M. 1996a. Managing insects, pp. 31–53. In: M.M. Peet (Ed.), *Sustainable Practices for Vegetable Production in the South*. Focus Publishing, R. Pullins Company, Newburyport, MA.

Peet, M.M. 1996b. Managing diseases, pp. 55–74. In: M.M. Peet (Ed.), *Sustainable Practices for Vegetable Production in the South*. Focus Publishing, R. Pullins Company, Newburyport, MA.

Peet, M.M. 1996c. Weed management pp. 79–88. In: M.M. Peet (Ed.), *Sustainable Practices for Vegetable Production in the South*. Focus Publishing, R. Pullins Company, Newburyport, MA.

Peet, M.M. 1996d. Tomato, pp. 149–157. In: M.M. Peet (Ed.), *Sustainable Practices for Vegetable Production in the South*. Focus Publishing, R. Pullins Company, Newburyport, MA.

Peet, M.M. 1997. Greenhouse crop stress management, pp. 91–101. In: R. Wignarajah (Ed.), *Proceedings 18th Annual Conference on Hydroponics*. Hydroponic Society of America, San Ramon, CA.

Peet, M.M. and M. Bartholomew. 1996. Effect of night temperature on pollen characteristics, growth, and fruit set in tomato. *J. Am. Soc. Hortic. Sci.* 121(34):514–519.

Peet, M.M. and D. Willits. 1995. Role of excess water in tomato fruit cracking. *HortScience* 30:65–68.

Pena, J.G. 1985. Economic considerations, marketing and financing of greenhouse vegetable production, pp. 77–87. In: A.J. Savage (Ed.), *Hydroponics Worldwide: State of the Art in Soilless Crop Production*. International Center for Special Studies, Honolulu, HI.

Peñalosa, J.M., M.J. Sarro, E. Revilla, R. Carpena, and C. Cadah'a. 1989. Influence of phosphorus supply on tomato plant nutrition. *J. Plant Nutr.* 12:647–657.

Picken, A.J.F., K. Stewart, and D. Klapwijk. 1986. Germination and vegetative growth, pp. 111–166. In: J.G. Atherton and J. Rudich (Eds.), *The Tomato Crop: A Scientific Basis for Improvement*. Chapman & Hall, New York.

Pierson, T. 1997. Consumers are in the drivers seat. *Am. Veg. Grower* 45(4):34.

Pill, W.G. and V.N. Lambeth. 1977. Effects of NH_4^+ and NO_3^- nutrition with and without pH adjustment on tomato growth, ion composition and water relations. *J. Am. Soc. Hortic. Sci.* 102:78–81.

Pill, W.G., V.N. Lambeth, and T.M. Hinckley. 1978. Effects of nitrogen form and level on ion concentration, water stress, and blossom-end rot incidence in tomato, *J. Am. Soc. Hortic. Sci.* 103(2):265–286.

Plummer, C. 1992. *United States Tomato Statistics, 1960–1990*. USDA Statistical Bulletin No. 841, U.S. Government Printing Office, Washington, D.C.

Polston, J.E. and P.K. Anderson. 1997. The explosion of whitefly-transmitted geminivirus in tomato in the Americas, pp. 38–42. In: C.S. Vavrina, P.R. Gilreath, and J.W. Noling (Eds.), *1977 Florida Tomato Institute Proceedings*. Citrus and Vegetable Magazine, University of Florida, Gainesville, FL.

Poncavage, J. 1997a. Grow great tasting early tomatoes. *Org. Gardening* 44(1):41–45.

Poncavage, J. 1997b. Timeless tomatoes. *Org. Gardening* 44(3):34–39.

Porter, J.R. and D.W. Lawlor (Eds.). 1991. *Plant Growth: Interactions with Nutrition and Environment*. Society for Experimental Biology Seminar Series 43, Cambridge University Press, Cambridge, London, England.

Powell, C.C. 1995. *Botrytis*: New management strategies for an old blight. *Nursery Manage. Prod.* 11:58–59.

Raloff, J. 1997. When tomatoes see red: The horticultural tricks colored mulch can play. *Sci. News* 152:376–377.

Ray, R.M. (Ed.). 1976. *All About Tomatoes*. 2nd ed. Ortho Books, Chevron Chemical Co., San Francisco, CA.

Raymond, D. and J. Raymond. 1978. *The Gardens for All Book of Tomatoes*. Gardens for All, Burlington, VT.

Resh, H.M. 1993. *Hydroponic Tomatoes for the Home Gardener*. Woodbridge Press, Santa Barbara, CA.

Resh, H.M. 1995. *Hydroponic Food Production*. 5th ed. Woodbridge Press, Santa Barbara, CA.

Reuter, D.J. and J.B. Robinson (Eds.). 1997. *Plant Analysis: An Interpretation Manual*. 2nd ed. CSIRO Publishing, Collingwood, Australia.

Roberts, W.J. and S. Kania. 1996. Screening for insect exclusion from greenhouses. In: *Proceedings 26th International Congress of American Society of Plasticulture*. Atlantic City, NJ.

Roberts, W.J. and D. Specca. 1997. The Burlington County research and development greenhouse, pp. 19–27. In: R. Wignarajah (Ed.), *Proceedings 18th Annual Conference on Hydroponics*. Hydroponic Society of America, San Ramon, CA.

Roorda van Eysinga, J.P.N.L. and K.W. Smilde. 1981. *Nutritional Disorders in Glasshouse Tomatoes, Cucumbers, and Lettuce*. Centre for Agricultural Publishing and Documentation, Wageningen, The Netherlands.

Rorabaugh, P.A. 1995. A brief and practical trek through the world of hydroponics, pp. 7–14. In: M. Bates (Ed.), *Proceedings of the 16th Annual Conference on Hydroponics*. Hydroponic Society of America, San Ramon, CA.

Rubatzky, V.E. and M. Yamaguchi. 1997. Tomato, pp. 533–552. In: V.E. Rubatzky and M. Yamaguchi (Eds.), *World Vegetables: Principles, Production, and Nutritive Values*. Chapman & Hall, New York.

Rudich, J. and U. Luchinisky. 1986. Water economy, pp. 335–367. In: J.G. Atherton and J. Rudich (Eds.), *The Tomato Crop: A Scientific Basis for Improvement*. Chapman & Hall, New York.

Rude, P.A. 1985. *IPM for Tomatoes*. University of California Publication 3274, Davis, CA.

Sargent, S.A., F.S. Maul, C.L. Moretti, and C.A. Sims. 1997. Harvest maturity, storage temperature and internal bruising affect tomato flavor, pp. 22–24. In: C.S. Vavrina, P.R. Gilreath, and J.W. Noling (Eds.), *1977 Florida Tomato Institute Proceedings*. Citrus and Vegetable Magazine, University of Florida, Gainesville, FL.

Savage, A.J. (Ed.). 1985. *Hydroponics Worldwide: State of the Art in Soilless Crop Production*. International Center for Special Studies, Honolulu, HI.

Savage, A.J. 1989. *Master Guide to Planning Profitable Hydroponic Greenhouse Operations*. Revised ed. International Center for Special Studies, Honolulu, HI.

Schales, F.D. 1985. Harvesting, packaging, storage, and shipping of greenhouse vegetables, pp. 70–76. In: A.J. Savage (Ed.), *Hydroponics Worldwide: State of the Art in Soilless Crop Production*. International Center for Special Studies, Honolulu, HI.

Schippers, P.A. 1979. *The Nutrient Flow Technique*. V.C. Mimeo 212, Department of Vegetable Crops, Cornell University, Ithaca, NY.

Schon, M. 1992. Tailoring nutrient solution to meet the demands of your plants, pp. 1–7. In: D. Schact (Ed.), *Proceedings of the 13th Annual Conference on Hydroponics*. Hydroponic Society of America, San Ramon, CA.

Schuster, D. 1997. Silver leaf whitefly threshold levels for irregular ripening in tomato, pp. 47–52. In: C.S. Vavrina, P.R. Gilreath, and J.W. Noling (Eds.), *1997 Florida Tomato Institute Proceedings*. Citrus and Vegetable Magazine, University of Florida, Gainesville, FL.

Schwarz, Meier. 1997. Carbon toxicity in plants, pp. 137–142. In: R. Wignarajah (Ed.), *Proceedings 18th Annual Conference on Hydroponics*. Hydroponic Society of America, San Ramon, CA.

Sheldrake, R. 1980. It's in the bag. *Greenhouse Grower* 28:33–34, 50.

Shipp, J.L., G.J. Boland, and L.A. Shaw. 1991. Integrated pest management of disease and anthropod pests of greenhouse vegetable crops in Ontario: Current status and future possibilities. *Can. J. Plant Sci.* 71:887–914.

Short, T.H., A. El-Attal, and H.M. Keener. 1997. *Decisions and Risk for Hydroponic Greenhouse Tomato Production*. Mimeograph. Ohio State University, Wooster, OH.

Slack, G. 1986. CO_2-enrichment of tomato crops, pp. 151–163. In: H.Z. Enoch and B.A. Kimball (Eds.), *Carbon Dioxide Enrichment of Greenhouse Crops*. Volume II: Physiology, Yield and Economics. CRC Press, Boca Raton, FL.

Smith, A. 1994. *The Tomato in America*. University of South Carolina Press, Columbia, SC.

Smith, T.N. 1993. How to use *E. Formosa* to control whiteflies. *Am. Veg. Grower* 41:47–48.

Snyder, R.G. 1993a. The U.S. greenhouse vegetable industry update and greenhouse vegetables in Mississippi. In: T.D. Carpenter and C.R. Gibbs (Eds.), *Proceedings American Greenhouse Vegetable Growers Conference and Trade Show*, August 19–21, Denver, CO.

Snyder, R.G. 1993b. *A Spreadsheet Approach to Fertilization Management for Greenhouse Tomatoes*. Mississippi Agricultural and Forestry Experiment Station Bulletin 1003, Mississippi State, MS.

Snyder, R.G. 1995. *Starting Vegetable Transplants*. Mississippi State Extension Service Publication 1995, Mississippi State, MS.

Snyder, R.G. 1996a. Greenhouse tomatoes—The basics of successful production, pp. 3–6. In: *Greenhouse Tomato Seminar*. ASHS Press, American Society for Horticultural Science, Alexandria, VA.

Snyder, R.G. 1996b. *Environmental Control for Greenhouse Tomatoes*. Mississippi State Extension Service Publication 1879, Mississippi State, MS.

Snyder, R.G. 1996c. *Fertigation: The Basics of Injecting Fertilizer for Field-Grown Tomatoes*. Mississippi State Extension Service Publication 2037, Mississippi State, MS.

Snyder, R.G. 1997a. *Greenhouse Tomato Handbook*. Mississippi State Extension Service Publication 1828, Mississippi State, MS.

Snyder, R.G. 1997b. *Greenhouse Tomato Pest Management in Mississippi*. Mississippi State Extension Service Publication 1861, Mississippi State, MS.

Snyder, R.G. and W.L. Bauerle. 1985. Water frequency and media volume affect growth, water status, yield, and quality of greenhouse tomatoes. *HortScience* 29(2):205–207.

Soffer, H. 1985. Israel: Current research and developments, pp. 123–130. In: A.J. Savage (Ed.), *Hydroponics Worldwide: State of the Art in Soilless Crop Production*. International Center for Special Studies, Honolulu, HI.

Soffer, H. 1988. Research on aero-hydroponics, pp. 69–74. In: *Proceedings of the 9th Annual Conference on Hydroponics*. Hydroponic Society of America, Concord, CA.

Sonneveld, C. 1985. *A Method for Calculating the Composition of Nutrient Solutions for Soilless Cultures*. No. 10, 2nd translated ed. Glasshouse Crops Research and Experiment Station, Naaldwijk, The Netherlands.

Sonneveld, C. 1989. Rockwool as a substrate in protected cultivation. *Chron. Hortic.* 29(3):33–36.

Stall, W.M. and J.P. Gilbreath. 1997. Weed control in tomato, pp. 88–91. In: C.S. Vavrina, P.R. Gilreath, and J.W. Noling (Eds.), *1997 Florida Tomato Institute Proceedings*. Citrus and Vegetable Magazine, University of Florida, Gainesville, FL.

Statistics Canada. 1993. *Greenhouse Industry*. Catalogue 22-202 Annual, Statistics Canada, Ottawa, Canada.

Stephens, D. 1997. Whitefly meets its match. *Ag Consultant* 53:36.

Stevens, M.A. 1986. The future of the field crop, pp. 559–579. In: J.G. Atherton and J. Rudich (Eds.), *The Tomato Crop: A Scientific Basis for Improvement*. Chapman & Hall, New York.

Stevens, M.A. and C.M. Rick. 1986. Genetics and breeding, pp. 35–109. In: J.G. Atherton and J. Rudich (Eds.), *The Tomato Crop: A Scientific Basis for Improvement*. Chapman & Hall, New York.

Stevens, M.A., A.A. Kader, M. Albright-Holton, and M. Algazi. 1977. Genotype variation for flavor and composition in fresh market tomatoes. *J. Am. Soc. Hortic. Sci.* 102:680–689.

Straver, W.A. 1996a. Inert growing media for greenhouse tomatoes, pp. 13–15. In: *Proceedings of Greenhouse Tomato Seminar*. ASHS Press, American Society for Horticultural Science, Alexandria, VA.

Straver, W.A. 1996b. Nutrition of greenhouse tomatoes on inert substrates in northern latitudes, pp. 31–33. In: *Proceedings of Greenhouse Tomato Seminar.* ASHS Press, American Society for Horticultural Science, Alexandria, VA.

Syltie, P.W., S.W. Melsted, and W.M. Walker. 1972. Rapid tissue tests as indicators of yield, plant composition, and fertility for corn and soybeans. *Commun. Soil Sci. Plant Anal.* 3:37–49.

Takahnashi, E., J.F. Ma, and Y. Miyake. 1990. The possibility of silicon as an essential element for higher plants, pp. 99–122. In: *Comments on Agriculture and Food Chemistry.* Gordon and Breach Science Publishers, London, Great Britain.

Taylor, I.B. 1986. Biosystematics of the tomato, pp. 1–34. In: J.G. Atherton and J. Rudich (Eds.), *The Tomato Crop: A Scientific Basis for Improvement.* Chapman & Hall, New York.

Taylor, J.D. 1983. *Grow More Nutritious Vegetables without Soil.* Parkside Press Publishing, Santa Ana, CA.

Thomas, L. 1995–1996. Hydroponic tomatoes: The flavor factor. *Growing Edge* 7(2):23–26.

Tindall, J.A., H.A. Mills, and D.E. Radcliffe. 1990. The effect of root zone temperature on nutrient uptake of tomato. *J. Plant Nutr.* 13:939–956.

Tite, R.L. 1983. *Growing Tomatoes: A Greenhouse Guide.* Primer 2, Ministry of Agriculture, Fisheries and Food, London, England.

Tonge, P. 1979. *The Good Green Garden.* Harpswell Press, Brunswick, ME.

Tonge, P. 1993. *Growing Your Garden the Earth-Friendly Way.* Christian Science Publishing Society, Boston, MA.

Tripp, K.E., M.M. Peet, D.M. Pharr, D.H. Willits, and P.V. Nelson. 1991. CO_2-enhanced yield and foliar deformation among tomato genotypes in elevated CO_2 environment. *Plant Physiol.* 96:713–719.

Trudel, M.J. and A. Gosselin. 1982. Influence of soil temperature in greenhouse tomato production. *HortScience* 17:928–929.

Tyler, K.B. and O.A. Lorenz. 1991. *Fertilizer Guide for California Vegetable Crops.* Department of Vegetable Crops Special Publication. University of California, Davis, CA.

USDA. 1997. *Agricultural Statistics, 1997.* USDA National Agricultural Service. United States Government Printing Office, Washington, D.C.

Valdez, R.B. 1978. Nematodes attacking tomato and their control, pp. 136–152. In: R. Cowell (Ed.), *1st International Symposium on Tropical Tomato.* Asian Vegetable Research and Development Center, Shanhua, Taiwan, Republic of China.

van de Vooren, J., G.W.H. Wells, and G. Hayman. 1986. Glasshouse crop production, pp. 581–623. In: J.G. Atherton and J. Rudich (Eds.), *The Tomato Crop: A Scientific Basis for Improvement.* Chapman & Hall, New York.

Van Patten, G.F. 1991. *Gardening: The Rockwool Book.* Van Patten Publishing, Portland, OR.

VanSickle, J.J. 1996. Florida tomatoes in a global market, pp. 1–6. In: C.S. Vavrina (Ed.), *1996 Proceedings of the Florida Tomato Institute.* PRO 108, University of Florida, Gainesville, FL.

Vavrina, C.S. and M.D. Orzolek. 1993. Tomato transplant age: A review. *HortTechnology* 3:313–316.

Vavrina, C.S., K. Armbrester, and M. Pena. 1997a. Heirloom tomato cultivar testing at the Southwest Research and Education Center, pp. 12–15. In: C.S. Vavrina, P.R. Gilreath, and J.W. Noling (Eds.), *1997 Florida Tomato Institute Proceedings.* Citrus and Vegetable Magazine, University of Florida, Gainesville, FL.

Vavrina, C.S., P.R. Gilreath, and J.W. Noling (Eds.). 1997b. *1997 Florida Tomato Institute Proceedings.* Citrus and Vegetable Magazine, University of Florida, Gainesville, FL.

Vavrina, C. 1991. Does tomato transplant age make a difference? *Vegetarian* 91–04:6.

Verwer, F.L. and J.J.C. Wellman. 1980. The possibilities of Grodan rockwool in horticulture, pp. 263–278. In: *Fifth International Congress on Soilless Culture*. International Society for Soilless Culture, Wageningen, The Netherlands.

Villareal, R.L. and S.H. Lai. 1978. Development of heat-tolerant tomato varieties in the tropics, pp. 188–200. In: R. Cowell (Ed.), *1st International Symposium on Tropical Tomato*. Asian Vegetable Research and Development Center, Shanhua, Taiwan, Republic of China.

Vitosh, M.L., D.D. Warncke, and R.E. Lucas. 1994. *Secondary and Micronutrients for Vegetables and Field Crops*. Michigan Extension Bulletin E-486, Michigan Agricultural Experiment Station, East Lansing, MI.

von Uexkull, H.R. 1979. Tomato: Nutritional and fertilizer requirements in the tropics, pp. 65–78. In: R. Cowell (Ed.), *1st International Symposium on Tropical Tomato*. AVRDC Publication 78-59, Asian Vegetable Research and Development Center, Shanhua, Taiwan, Republic of China.

Voogt, W. 1993. Potassium and Ca ratios of year round tomato crops. *Acta Hortic*. 339:99–112.

Voogt, W. and C. Sonneveld. 1997. Nutrient management in closed growing systems for greenhouse production, pp. 83–102. In: E. Goto, K. Kurate, M. Hayashi, and S. Sase (Eds.), *Plant Production in Closed Ecosystems*. Kluwer Academic Publishers, Dordrecht, The Netherlands.

Wallace, A. 1983. The third decade of synthetic chelating agents in plant nutrition. *J. Plant Nutr*. 6:425–562.

Walls, J. 1990. *Tons of Tomatoes* (No. 812 video). The New Garden, San Antonio, TX.

Wang, J.-Y. 1963. *Agricultural Meteorology*. Pacemaker Press, Milwaukee, WI.

Ward, G.M. 1964. Greenhouse nutrition in a growth analysis study. *Plant Soil* 21:125–133.

Ward, G.M. 1977. Manganese deficiency and toxicity in tomatoes. *Can. J. Plant Sci*. 57:107–115.

Waterman, M.P. 1993–1994. Building a better tomato: Research review. *Growing Edge* 5(2):20–25, 69.

Waterman, M.P. 1996. The good, bad, and the ugly: Insect research doesn't discriminate. *Growing Edge* 7(3):22–28, 75.

Watterson, J.C. 1986. Weed control, pp. 485–510. In: J.G. Atherton and J. Rudich (Eds.), *The Tomato Crop: A Scientific Basis for Improvement*. Chapman & Hall, New York.

Weaver, W.W. 1998. A cook's favorite heirloom tomatoes. *Org. Gardening* 45(5):38–42.

Weingartner, D.P. 1997. Late blight status in Florida potatoes and tomatoes, pp. 43–46. In: C.S. Vavrina, P.R. Gilreath, and J.W. Noling (Eds.), *1997 Florida Tomato Institute Proceedings*. Citrus and Vegetable Magazine, University of Florida, Gainesville, FL.

Weir, R.G. and G.C. Cresswell. 1993. *Plant Nutritional Disorders. Vegetable Crops*. Volume 3, Florida Science Source, Lake Alfred, FL.

Westerman, R.L. (Ed.). 1990. *Soil Testing and Plant Analysis*. 3rd ed. SSSA Book Series No. 3, Soil Science Society of America, Madison, WI.

Wignarajah, K. (Ed.). 1997. *Proceedings 18th Annual Conference on Hydroponics*. Hydroponic Society of America, San Ramon, CA.

Wilcox, G.E. 1991. Nutrient control in hydroponic systems, pp. 50–53. In. S. Knight (Ed.), *Proceedings of the 12th Annual Conference on Hydroponics*. Hydroponic Society of America, San Ramon, CA.

Wilcox, G.E., J.E. Hoff, and C.M. Jones. 1973. Ammonium reduction of calcium and magnesium content of tomato and seed corn leaf tissue and influence on incidence of blossom-end rot of tomato fruit. *J. Am. Soc. Hortic. Sci*. 98(1):86–89.

Willits, D.H. and M.M. Peet. 1989. Predicting yield responses to different greenhouse CO_2 enrichment scheme: Cucumbers and tomatoes. *Agric. Meteorol*. 44:275–293.

Winsor, G. and P. Adams. 1987. *Diagnosis of Mineral Disorders in Plants: Glasshouse Crops.* Volume 3. Chemical Publishing, New York.

Wittwer, S.H. 1993. World-wide use of plastics in horticultural production. *HortTechnology* 3:6–19.

Wittwer, S.H. and N. Castilla. 1995. Protected cultivation of horticultural crops worldwide. *HortTechnol.* 5:6–23.

Wittwer, S.H. and D. Honma. 1969. *Greenhouse Tomatoes: Guidelines for Successful Production.* Michigan State University Press, East Lansing, MI.

Yang, C.T. 1978. Bacterial and fungal diseases of tomato, pp. 111–123. In: R. Cowell (Ed.), *1st International Symposium on Tropical Tomato.* Asian Vegetable Research and Development Center, Shanhua, Taiwan, Republic of China.

Appendix I: Reference Books and Videos

BOOKS

Abeles, F.B., P.W. Morgan, and M.E. Saltveit. 1992. *Ethylene in Plant Biology*. Academic Press, San Diego, CA.

Adams, C.F. and M. Richardson. 1977. *Nutritive Value of Foods*. USDA-ARS Home and Garden Bulletin Number 72, U.S. Government Printing Office, Washington, D.C.

Anon. 1998. *Soil and Plant Analysis Laboratory Registry for the United States and Canada*. CRC Press, Boca Raton, FL.

Anon. 1990. *Integrated Pest Management—Tomatoes*. 3rd ed. Publication 3274, Division of Agriculture and Natural Resources, University of California, Oakland, CA.

Anon. 1996. *Proceedings of Greenhouse Tomato Seminar*. ASHS Press, American Society for Horticultural Science, Alexandria, VA.

Anon. 1998. *Farm Chemical Handbook '98*. Meister Publishing, Willoughby, OH.

Atherton, J.G. and J. Rudich (Eds.). 1986. *The Tomato Crop: A Scientific Basis for Improvement*. Chapman & Hall, New York.

Ball, V. (Ed.). 1985. *Ball Red Book: Greenhouse Growing*. 14th ed. Reston Publishing, Reston, VA.

Barber, S.A. 1995. *Soil Nutrient Bioavailability: A Mechanistic Approach*. 2nd ed. John Wiley & Sons, New York.

Barber, S.A. and D.R. Bouldin (Eds.). 1984. *Roots, Nutrients and Water Influx, and Plant Growth*. ASA Publication 136, American Society of Agronomy, Madison, WI.

Bartholomew, M. 1981. *Square Foot Gardening*. Rodale Press, Emmaus, PA.

Bennett, W.F. (Ed.). 1993. *Nutrient Deficiencies and Toxicities in Crop Plants*. APS Press, American Phytopathological Society, St. Paul, MN.

Bergmann, W. 1992. *Nutritional Disorders of Plants: Developments, Visual and Analytical Diagnosis*. Gustav Fischer Verlag, Jenna, Germany.

Boodley, J.W. and R. Sheldrake, Jr. 1972. *Cornell Peat-Lite Mixes for Commercial Plant Growing*. Information Bulletin 43, Cornell University, Ithaca, NY.

Brooks, W.M. 1969. *Growing Greenhouse Tomatoes*. SB-19. Cooperative Extension Service, Ohio State University, Columbus, OH.

Bunt, A.C. 1988. *Media and Mixes for Container-Grown Plants*. 2nd ed. Unwin Hyman, London, England.

Carson, E.W. (Ed.). 1974. *The Plant Root and Its Environment*. University Press of Virginia, Charlottesville, VA.

Cowell, R. (Ed.). 1978. *1st International Symposium on Tropical Tomato*. Asian Vegetable Research and Development Center, Shanhua, Taiwan, Republic of China.

Epstein, E. 1972. *Mineral Nutrition of Plants: Principles and Perspectives*. John Wiley & Sons, New York.

Giacomelli, G.A. and K.C. Ting. 1994. *Greenhouse Systems: Automation, Culture and Environment.* Northeast Regional Agricultural Engineering Service Publication NRAES-72, NRAES, Cornell University, Ithaca, NY.

Glass, A.D.M. 1989. *Plant Nutrition: An Introduction to Current Concepts.* Jones & Bartlett Publishers, Boston, MA.

Goto, E., K. Kurate, M. Hayashi, and S. Sase (Eds.). 1997. *Plant Production in Closed Ecosystems.* Kluwer Academic Publishers, Dordrecht, The Netherlands.

Halliday, D.J. and M.E. Trenkel (Eds.). 1992. *IFA World Fertilizer Use Manual*, pp. 289–290, 331–337. International Fertilizer Industry Association, Paris, France.

Hendrickson, R. 1977. *The Great American Tomato Book.* Doubleday, Garden City, NY.

Hendrickson, R. 1997. *Tomatoes.* A Burpee Book. Macmillan (Simon & Schuster), New York.

Hessayon, D.G. 1997. *The New Vegetable and Herb Expert.* Expert Books, Transworld Publishers, London, England.

Hoagland, D.R. and D.I. Arnon. 1950. *The Water Culture Method for Growing Plants without Soil.* Circular 347, California Agricultural Experiment Station, University of California, Berkeley, CA.

Hochmuth, G. (Ed.). 1991. *Florida Greenhouse Vegetable Protection Handbook.* Volume 3. Greenhouse Vegetable Crop Protection Guide. Circular CP48, University of Florida, Institute of Food and Agricultural Sciences, Gainesville, FL.

Hussey, N.W. and N. Scopes (Eds.). 1985. *Biological Pest Control: In Greenhouse Experience,* Blandford Press, Poole, Dorset, United Kingdom.

James, H. 1996. *The Farmer's Guide to the Internet.* TVA Rural Studies, University of Kentucky, Lexington, KY.

Johnson, D.B. and E.H. Brindle. 1976. *Vegetable Gardening Basics.* Burgess Publishing, Broken Arrow, OK.

Jones, J.B., Jr. 1997a. *Plant Nutrition Manual.* CRC Press, Boca Raton, FL.

Jones, J.B., Jr. 1997b. *Hydroponics: A Practical Guide for the Soilless Grower.* St. Lucie Press, Boca Raton, FL.

Kabata-Pendias, A. and H. Pendias. 1994. *Trace Elements in Soils and Plants.* 2nd ed. CRC Press, Boca Raton, FL.

Kader, A.A. (Ed.). 1992. *Postharvest Technology of Horticultural Crops of California.* California Division of Agriculture and Natural Resources Publication 3311, Sacramento, CA.

Kalra, Y.P. (Ed.). 1998. *Handbook on Reference Methods for Plant Analysis.* CRC Press, Boca Raton, FL.

Keller, J. and R.D. Bliesner. 1990. *Sprinkle and Trickle Irrigation.* Van Nostrand Reinhold, New York.

Lindsay, W.L. 1979. *Chemical Equilibria in Soils.* John Wiley & Sons, New York.

Lorenz, O.A. and D.N. Maynard. 1988. *Knott's Handbook for Vegetable Growers.* 3rd ed. John Wiley & Sons, New York.

Maas, E.F. and R.M. Adamson. 1971. *Soilless Culture of Commercial Greenhouse Tomatoes.* Publication 1460, Information Director, Canada Department of Agriculture, Ottawa, Ontario, Canada.

Malais, M. and W.J. Ravensberg. 1992. *Knowing and Recognizing the Biology of Glasshouse Pests and Their Natural Enemies.* Koppert Biological Systems. Koppert B.V. Berkel en Rodenrijs, The Netherlands.

Marschner, H. 1986. *Mineral Nutrition of Higher Plants.* Academic Press, New York.

Martin-Prével, P., J. Gagnard, and P. Gautier. 1984. *Plant Analysis: A Guide to the Nutrient Requirements of Temperate and Tropical Crops.* Lavoisier Publishing, New York.

Maynard, D.H. and G.J. Hochmuth. 1997. *Knott's Handbook for Vegetable Growers.* 4th ed. John Wiley & Sons, New York.

Mengel, K. and E.A. Kirkby. 1987. *Principles of Plant Nutrition,* 4th ed. International Potash Institute, Bern, Switzerland.
Mills, H.A. and J. B. Jones, Jr. 1996. *Plant Nutrition Manual II.* Micro-Macro Publishing, Athens, GA.
Mittleider, J.R. 1975. *More Food from Your Garden.* Woodbridge Press, Santa Barbara, CA.
Molyneux, C.J. 1988. *A Practical Guide to NFT.* T. Snap & Co., Preston, Lancashire, England.
Mortvedt, J.J. 1991. *Micronutrients in Agriculture.* 2nd ed. SSSA Book Series No. 4, Soil Science Society of America, Madison, WI.
Muckle, M.E. 1990. *Hydroponic Nutrients—Easy Ways to Make Your Own.* Revised ed. Growers Press, Princeton, B.C., Canada.
Nakayama, F.S. and D.A. Bucks. 1986. *Trickle Irrigation for Crop Production.* Elsevier Science Publishers, Amsterdam, The Netherlands.
Pais, I. and J.B. Jones, Jr. 1997. *The Handbook on Trace Elements.* St. Lucie Press, Boca Raton, FL.
Papadopoulos, A.P. 1991. *Growing Greenhouse Tomatoes in Soil and in Soilless Media.* Agriculture Canada Publication 186/E, Communications Branch, Agriculture Canada, Ottawa, Canada.
Parker, D. (Ed.). 1994. *The Best of the Growing Edge.* New Moon Publishing, Corvallis, OR.
Parnes, R. 1990. *Fertile Soil, A Grower's Guide to Organic and Inorganic Fertilizers.* AgAccess, Davis, CA.
Peet, M.M. (Ed.). 1996. *Sustainable Practices for Vegetable Production in the South.* Focus Publishing, R. Pullins Company, Newburyport, MA.
Porter, J.R. and D.W. Lawlor (Eds.). 1991. *Plant Growth: Interactions with Nutrition and Environment.* Society for Experimental Biology Seminar Series 43, Cambridge University Press, Cambridge, London, England.
Puma, J. 1985. *The Complete Urban Gardener.* Harper & Row Publishers, New York.
Ray, R.M. (Ed.). 1976. *All About Tomatoes.* 2nd ed. Ortho Books, Chevron Chemical Co., San Francisco, CA.
Raymond, D. and J. Raymond. 1978. *The Gardens for All Book of Tomatoes.* Gardens for All, Burlington, VT.
Resh, H.M. 1993. *Hydroponic Tomatoes for the Home Gardener.* Woodbridge Press, Santa Barbara, CA.
Resh, H.M. 1995. *Hydroponic Food Production.* 5th ed. Woodbridge Press, Santa Barbara, CA.
Reuter, D.J. and J.B. Robinson (Eds.). 1997. *Plant Analysis: An Interpretation Manual.* 2nd ed. CSIRO Publishing, Collingwood, Australia.
Roorda van Eysinga, J.P.N.L. and K.W. Smilde. 1981. *Nutritional Disorders in Glasshouse Tomatoes, Cucumbers, and Lettuce.* Centre for Agricultural Publishing and Documentation, Wageningen, The Netherlands.
Ross, J. 1998. *Hydroponic Tomato Production: A Practical Guide to Growing Tomatoes in Containers.* Casper Publications, Narabeen, Australia.
Rubatzk, V.E. and M. Yamaguchi (Eds.). 1997. *World Vegetables: Principles, Production, and Nutritive Values.* Chapman & Hall, New York.
Salunkhe, D.K. and S.S. Kadam (Eds.). 1998. *Handbook of Vegetable Science and Technology, Production, Composition, Storage, and Processing.* Marcel Dekker, New York.
Savage, A.J. (Ed.). 1985. *Hydroponics Worldwide: State of the Art in Soilless Crop Production.* International Center for Special Studies, Honolulu, HI.
Savage, A.J. 1989. *Master Guide to Planning Profitable Hydroponic Greenhouse Operations.* Revised ed. International Center for Special Studies, Honolulu, HI.
Selsam, M.E. 1970. *The Tomato and Other Fruit Vegetables.* William Morrow, New York.

Smith, D.L. 1987. *Rock Wool in Horticulture*. Grower Books, London, England.

Snyder, R.G. 1993b. *A Spreadsheet Approach to Fertilization Management for Greenhouse Tomatoes*. Mississippi Agricultural and Forestry Experiment Station Bulletin 1003, Mississippi State, MS.

Snyder, R.G. 1995. *Starting Vegetable Transplants*. Mississippi State Extension Service Publication 1995, Mississippi State, MS.

Snyder, R.G. 1996b. *Environmental Control for Greenhouse Tomatoes*. Mississippi State Extension Service Publication 1879, Mississippi State, MS.

Snyder, R.G. 1996c. *Fertigation: The Basics of Injecting Fertilizer for Field-Grown Tomatoes*. Mississippi State Extension Service Publication 2037, Mississippi State, MS.

Snyder, R.G. 1997a. *Greenhouse Tomato Handbook*. Mississippi State Extension Service Publication 1828, Mississippi State, MS.

Snyder, R.G. 1997b. *Greenhouse Tomato Pest Management in Mississippi*. Mississippi State Extension Service Publication 1861, Mississippi State, MS.

Taylor, J.D. 1983. *Grow More Nutritious Vegetables without Soil*. Parkside Press Publishing, Santa Ana, CA.

Tite, R.L. 1983. *Growing Tomatoes: A Greenhouse Guide*. Primer 2, Ministry of Agriculture, Fisheries and Food, London, England.

Tonge, P. 1979. *The Good Green Garden*. Harpswell Press, Brunswick, ME.

Tonge, P. 1993. *Growing Your Garden the Earth-Friendly Way*. Christian Science Publishing Society, Boston, MA.

Van Patten, G.F. 1991. *Gardening: The Rockwool Book*. Van Patten Publishing, Portland, OR.

Vavrina, C.S. (Ed.). 1996. *1996 Proceedings of the Florida Tomato Institute*. PRO 108, University of Florida, Gainesville, FL.

Vavrina, C.S., P.R. Gilreath, and J.W. Noling (Eds.). 1997. *1997 Florida Tomato Institute Proceedings*. Citrus and Vegetable Magazine, University of Florida, Gainesville, FL.

Vitosh, M.L., D.D. Warncke, and R.E. Lucas. 1994. *Secondary and Micronutrients for Vegetables and Field Crops*. Michigan Extension Bulletin E-486, Michigan Agricultural Experiment Station, East Lansing, MI.

Weir, R.G. and G.C. Cresswell. 1993. *Plant Nutritional Disorders. Vegetable Crops*. Volume 3, Florida Science Source, Lake Alfred, FL.

Westerman, R.L. (Ed.). 1990. *Soil Testing and Plant Analysis*. 3rd ed. SSSA Book Series No. 3, Soil Science Society of America, Madison, WI.

Wien, H.C. (Ed.). 1997. *The Physiology of Vegetable Crops*. Commonwealth Agricultural Bureau (CAB) International. Wallingford, United Kingdom.

Winsor, G. and P. Adams. 1987. *Diagnosis of Mineral Disorders in Plants: Glasshouse Crops*. Volume 3, Chemical Publishing, New York.

VIDEOS

Growing Training Workshop on Video. 1995. CropKing, 5050 Greenwich Road, Seville, OH 44273.

Nutrient Element Deficiencies in Tomato. 1993. St. Lucie Press, 2000 Corporate Blvd., NW, Boca Raton, FL 33431.

Plant Nutrition Basics. 1993. St. Lucie Press, 2000 Corporate Blvd., NW, Boca Raton, FL 33431.

Plant Analysis. 1993. St. Lucie Press, 2000 Corporate Blvd., NW, Boca Raton, FL 33431.

Tissue Testing. 1993. St. Lucie Press, 2000 Corporate Blvd., NW, Boca Raton, FL 33431.

Tons of Tomatoes (No. 812). 1990. The New Garden, San Antonio, TX.

Appendix II: Glossary

acre: a unit of square measure equal to 43,560 ft^2, or an area of a 208 ft square.
adventitious roots: roots that develop from the main stem or from the stem of a plant cutting.
aerobic: having oxygen (O_2) as a part of the environment.
aeroponics: a technique for growing plants hydroponically where the plant roots are suspended in a container and are either continuously or periodically bathed in a fine mist of nutrient solution.
alkaline soil: a soil having an alkaline, or basic reaction, that is, a pH above 7.0.
anaerobic: an environmental condition in which molecular oxygen (O_2) is deficient for chemical, physical, or biological processes.
anion: an ion carrying a negative charge, examples being borate (BO_3^{3-}), chloride (Cl^-), dihydrogen phosphate ($H_2PO_4^-$), monohydrogen phosphate (HPO_4^{2-}), nitrate (NO_3^-), and sulfate (SO_4^{2-}).
annual: one season's duration completed life cycle from germination to maturity and death.
available water: portion of water in soil that can be readily absorbed by roots; that soil moisture held in the soil between field capacity and permanent wilting percentage.
axil: the angle between the shoot and a leaf petiole, branch, etc.
beneficial elements: elements not essential for plants but which, when present, enhance plant growth.
berry: fleshy, many-seeded fruit with single or multiple carpels.
biological control: using natural forces, such as predators, to control harmful pests.
blossom-end rot (BER): the breakdown near the blossom end of fruit caused by imbalances in moisture (or other stress) or calcium, or both.
botrytis: a fungus that causes numerous diseases, such as gray mold in tomato, and causes several rots of fruit.
calcareous soil: soil having a pH greater than 7.0 due to the presence of free calcium carbonate ($CaCO_3$).
calyx: collective term of the sepals of a flower.
carotene: a yellow pigment and precursor of vitamin A.
cation: an ion carrying a positive charge, examples being ammonium (NH_4^+), calcium (Ca^{2+}), copper (Cu^{2+}), iron (Fe^{2+} or Fe^{3+}), hydrogen (H^+), potassium (K^+), magnesium (Mg^{2+}), manganese (Mn^{2+}), and zinc (Zn^{2+}).
cellulose: a structural material in the cell walls of plants.
chelates: a type of chemical compound in which a metallic atom (such as iron) is firmly combined with a molecule by means of multiple chemical bonds; the term which refers to the claw of a crab, illustrative of the way in which the atom is held.
chlorophyll: green pigment in chloroplasts necessary for photosynthesis.
chlorosis: condition whereby a plant or plant part is light green or greenish yellow because of poor chlorophyll development or destruction of chlorophyll.

cluster: refers to a group of tomato fruit growing from the same stem (*see* **truss**).
compound leaf: a leaf whose blade is divided into a number of distinct leaflets.
compost: rotted remains of organic debris.
controlled atmosphere: storage in which the atmospheric content is regulated.
cultivar: a horticultural variety or race that originated under cultivation and not essentially referable to botanical species; abbreviated as cv.
day-neutral plants: plants that are not affected by length of day or dark period with regard to floral initiation.
deficiency: describes the condition when an essential element is not in sufficient supply or proper form to adequately supply the plant or is not in sufficient concentration in the plant to meet the plant's physiological requirement; plants which therefore grow poorly and show visual signs of abnormality in color and structure.
degree-day: a unit of heat representing one degree above a given average daily base value, usually the minimum temperature for growth of the plant.
desucker: removal of undesirable shoots from a plant.
determinate: plant growth in which the shoot terminates in an inflorescence and further growth is arrested.
drip irrigation (trickle irrigation): method whereby water is applied slowly through small-orificed emitters.
electrical conductivity (EC): a measure of the electrical resistance of water, a nutrient solution, or a soil or medium solution, used to determine the level of ions in solution and as a means to determine potential effect on plant growth.
essential elements: elements, 16 in number, that are necessary for higher plants to complete their life cycle based on the criteria established by Arnon and Stout (*see* Chapter 4).
ethylene: gas (C_2H_4) having growth regulating capabilities; induces physiological responses; produced by plant tissues, especially by many fruit; hastens fruit ripening and abscission.
evapotranspiration: the total loss of water by evaporation from a given area of the soil surface and plant transpiration.
family: category of classification above genus and below order; suffix of family name usually "aceae."
fertigation: application of fertilizer through irrigation water.
fibrous roots: a root system in which both main and lateral roots have similar diameters.
flower: reproductive organ of a seed-bearing plant.
foliar feeding: application of nutrients to the foliage of plants.
fruit: a ripened ovary usually containing seeds and accessory parts; a seed-bearing structure usually eaten raw.
fruit set: the swelling and initial development of the ovary into a fruit.
genotype: genetic constitution, latent or expressed, of an organism.
genus: a group of closely related species clearly different from other groups.
germination: the beginning of growth in a seed.
greenhouse (glasshouse): a structure covered with a transparent material for the purpose of admitting natural light and used for growing plants.
hard pan: a hard layer of soil beneath the tilled zone through which water and root penetration are difficult.
hardening: procedure to acclimate plants to adverse environmental conditions (e.g., low temperature or low moisture).
hectare: a unit of square measure, 10,000 square meters (m^2), equivalent to 2.471 acreas (A).
herbicide: a chemical that kills plants.
hybrid: a cross between parents that are genetically unalike.

hydroponics: a word coined in the early 1930s by Dr. W.F. Gericke (a University of California researcher) to describe a soilless technique for growing plants; word which was derived from two Greek words: *hydro*, meaning water, and *ponics*, meaning labor—literally *working water;* hydroponics defined as the science of growing plants without the use of soil, but rather by use of an inert medium to which a nutrient solution containing those essential elements required by the plant for normal growth is applied to the roots.

indeterminate: shoot axis remains vegetative; does not terminate with an inflorescence.

inflorescence: an axis bearing flowers, or a flower cluster.

internode: region between nodes.

insecticide: a chemical substance that kills insects.

leaching: removal of soluble salts by the downward movement of water through a rooting medium.

leaf area index (LAI): leaf foliage density expressed as leaf area subtended per unit area or land.

lux: unit of light intensity.

lycopene: a carotenoid pigment; has no provitamin A value.

macronutrient: essential element required by plants in relatively large amounts; the elements being carbon (C), calcium (Ca), hydrogen (H), magnesium (Mg), nitrogen (N), oxygen (O), phosphorus (P), potassium (K), and sulfur (S).

mass flow: the movement of ions as a result of the flow of water; the ions are carried in the moving water.

micronutrient: chemical elements necessary in extremely small amounts for the growth of plants; the elements being boron (B), chloride (Cl), copper (Cu), iron (Fe), manganese (Mn), molybdenum (Mo), and zinc (Zn).

medium: material in and on which plants are grown (plural, **media**).

necrosis: plant tissue turning black due to disintegration or death of cells, usually caused by disease.

nematode: microscopic, nonsegmented roundworms that often cause or transmit diseases.

node: enlarged region of a stem that is generally solid where a leaf is attached and buds are located.

nutrient film technique (NFT): a technique for growing plants hydroponically in which the plant roots are suspended in a slow-moving stream of nutrient solution. This technique was developed by Dr. Allen Cooper in the 1970s.

ovary: swollen base of the pistil containing the ovules, which on fertilization develops into a fruit.

peat moss: commonly, the dried, shredded peat from sphagnum moss.

perlite: an aluminosilicate of volcanic origin; when this natural substance crushed and heated rapidly to 1000ºC, forms a white, lightweight aggregate with a closed cellular structure; has an average density of 8 lb f^{-3} (128 kg m^{-3}), and virtually no cation exchange capacity; is devoid of plant nutrients, contains some fluoride (17 mg F kg^{-1}), and is graded into various particle sizes for use as a rooting medium or an addition to soilless mixes.

pesticide: a general term that describes a material used to control any sort of pest.

petiole: stalk or stemlike structure of a leaf.

pH: negative log of hydrogen ion concentration of a soil; pH 7.0 denoting neutrality, < 7.0 denoting acidity, > 7.0 denoting alkalinity.

phloem: conductive tissue that transports synthesized substances to other plant parts.

photoperiod: relative length of period of light and darkness.

physiological disorder: disorder not caused by pathogens, but instead due to physiological dysfunction of the organism.

photosynthesis: process in which carbon dioxide (CO_2) and water (H_2O) in the presence of light are combined in chlorophyllous tissues to form carbohydrates and oxygen (O_2).

pine bark: a by-product of the processing of pine, usually southern yellow pine, for lumber; bark stripped from the tree being allowed to age in the natural environment for 6 months to 2 years, passed through a hammer mill, and then passed through a 1-in. screen; resulting material screened into fractions of various sizes for addition to organic mixes; has substantial cation exchange and water-holding capacities, and provides some degree of root disease control.

pistillate: having female flower.

plant analysis: a method of determining the total elemental content of whole plant or one of its parts and then relating the concentration found to the well-being of the plant in terms of its elemental requirements.

pollen: small, usually yellow, grains released from the anthers that carry the male gametes for fertilization.

polyethylene: a clear plastic material used as a covering for greenhouses; black polyethylene often used as a cover for a soil bench.

relative humidity: ratio of actual amount of water in air to the maximum amount (saturation) that air can hold at the same temperature expressed in percentage.

respiration: biological oxidation of organic matter by enzymes to obtain energy.

ripening: chemical and physical changes in a fruit that follow maturation.

rockwool: an inert fibrous material produced from a mixture of volcanic rock, limestone, and coke that is melted at 1500–2000°C, extruded as fine fibers, and then pressed into loosely woven sheets; has excellent water-holding capacity; for growing uses, rockwool sheets formed into slabs or cubes.

root: a part of a plant, usually underground, the main functions of which are to anchor the plant, and absorb moisture and nutrients.

scorch: leaf injury on the margins caused by a nutrient element stress, deficient or excess, or other forms of plant stress.

seed: the mature ovule of a flowering plant containing an embryo, a food supply, and a seed coat.

sepal: an outermost, often leaflike, nonsexual portion of a flower; part of a calyx.

senescence: a physiological aging process in which tissues in an organism deteriorate and finally die.

short-day plants: plants that flower when the dark period exceeds some critical length.

soil solution: the soluble materials in a soil, or soilless media, held in solution.

soluble salts: a measure of the concentration of ions in water (or nutrient solution) used to determine the quality of the water or solution, measuring in terms of its electrical conductivity.

species: a group of similar organisms capable of interbreeding and more or less distinctly different in morphological characteristics from other species in the same genus.

starch: complex polysaccharides of glucose; the form of food commonly stored by plants.

stem: the stalk of a plant or plant part.

stomate: having a very small opening (**stoma**, plural **stomata**) in the surface of a leaf for the exchange of gases and moisture.

sucker: adventitious shoot from the lower part of the plant; a stem that grows from the axil.

sunscald: injury caused by direct exposure to intense sunlight.

taproot: the main descending root of a plant.

thinning: removal of young plants to provide remaining plants more space to develop.

tissue: a group of cells of similar structure that perform a specific function.

tissue testing: a method for determining the concentration of the soluble form of an element in the plant by analyzing sap that has been physically extracted from a particular plant part, usually from stems or petioles: tests usually limited to the determination of nitrate (NO_3), phosphate (P), potassium (K), and iron (Fe); normally performed using simple analysis kits; and the elemental concentration form being related to the well-being of the sampled plant.

toxicity: the condition in which an element is sufficiently in excess in the rooting medium, nutrient solution, or plant to be detrimental to the plant's normal growth and development.

transpiration: water loss from plant tissues, usually leaves.

trickle irrigation: *see* **drip irrigation**.

truss: refers to a group of tomato fruit growing from the same stem; *see* **cluster**.

turgor: normal inflation of cells due to internal pressure on the cell walls, usually pressure exerted by water.

variety (botanical): a subdivision of a species with distinct morphological characters given a Latin binomial name according to rules of the International Code of Botanical Nomenclature.

vegetable: food plant, most often herbaceous annual, cultivated or gathered; possible for edible portions to be roots, stems, leaves, floral parts, fruits, and seeds; usually high in water content; eaten raw or cooked.

vermiculite: magnesia mica heated to expand to many times its original volume; an ingredient added to some forms of soilless mixes.

wilt: condition in which the plant droops because of a decrease in cell turgor due to the lack of water.

xylem: plant conductive tissues that transport water and absorbed nutrients from roots to other tissues.

Appendix III: Characteristics of the Six Major Essential Elements and the Seven Micronutrients Related to Tomato Plant/Fruit Production

THE MAJOR ELEMENTS

Nitrogen (N)

Atomic Number: 7 **Atomic Weight:** 14.00

Discoverer of Essentiality and Year: DeSaussure—1804

Designated Element: major element

Function: used by plants to synthesize amino acids and form proteins, nucleic acids, alkaloids, chlorophyll, purine bases, and enzymes

Mobility in the Tomato Plant: mobile

Requirement Level: medium

Forms Utilized by the Tomato Plant: ammonium (NH_4^+) cation and nitrate (NO_3^-) anion

Common Fertilizers:

Sources	Formula	N (Other) (%)
	Inorganic	
Ammonium dihydrogen phosphate	[$NH_4H_2PO_4$]	11 (21 P)
Ammonium nitrate	[NH_4NO_3]	34 (17 NH_4, 17 NO_3)
Ammonium sulfate	[$(NH_4)_2SO_4$]	21 (24 S)
Diammonium hydrogen phosphate	[$(NH_4)_2HPO_4$]	16–18 (21 P)

Sources	Formula	N (Other) (%)
	Inorganic	
Calcium nitrate	[$Ca(NO_3)_2 \cdot 4H_2O$]	16 (19 Ca)
Potassium nitrate	[KNO_3]	13 (36 K)
Urea	$CO(NH_2)_2$	45–46
Sulfur-coated urea	$CO(NH_2)_2$–S	40
Urea–formaldehyde	$CO(NH_2)_2$–CH_2O	38
	Organic	
Cottonseed meal		12–13
Milorganite		12.0
Animal manure		10–12
Sewage sludge		10–20
Chicken litter		20–40

Concentration in Nutrient Solution: 100–200 mg L^{-1} (ppm) of N [as either the ammonium (NH_4^+) cation or nitrate (NO_3^-) anion]; 3–4 parts as the nitrate (NO_3^-) anion

Fertilizer Nitrogen Recommendation: range: 50–220 lb A^{-1} of N (56–246 kg ha^{-1} of N); mean: 110 lb A^{-1} of N (123 kg ha^{-1} of N)

Typical Deficiency Symptoms: growth slowing and plants becoming stunted and light green in color, and as the deficiency advances, the lower or older leaves yellowing and dying; plants maturing early; fruit yield and quality declining

Symptoms of Excess: plants dark green in color and easily susceptible to disease and insect attack and moisture stress; blossom absorption occurring; fruit set and quality declining

Ammonium Toxicity: cupping or slight rolling of the older leaves occurring, lesions on stems and leaves occurring, blossom-end rot (BER) occurring on the fruit; and as the plant matures, the vascular tissue at the base of the plant beginning to deteriorate (cutting the stem at the base of the plant causing discoloration of the vascular tissue due to decay) with wilting occurring during periods of high atmospheric demand, followed by eventual death of the plant

Sufficiency Range in the Tomato Plant:

Tissue	Plant Stage	N Content (%)	Source
	Total N		
Compound leaves adjacent to top inflorescences	Midbloom	4.0–6.0	Mills and Jones, 1996
Mature leaves from new growth (greenhouse)	Mature plants	2.8–4.2	
Whole leaf	Prior to fruiting	4.0–5.0	Hochmuth, 1996b
	During fruiting	3.5–4.0	
Mature leaf (healthy plant)	Not given	2.8–4.9	Halliday and Trenkel, 1992

Tissue	Plant Stage	N Content (%)	Source
		Total N	
Young mature leaf (optimum fertility)	Half fruit	2.7	
Youngest open leaf blade (adequate)	Early flower	5.0–6.0 (Critical value: 4.9)	Reuter and Robinson, 1997
	Early fruit set	4.6–6.0	
	First mature fruit	4.5–4.6 (Critical value: 4.45)	
	Midharvest	4.5–5.5	
		Extractable N	
Nitrate–Nitrogen (NO_3–N) [mg kg^{-1} (ppm)]			
Plant stems	Very early growth	<100 Deficient	Beverly, 1994
		500–600 Adequate	
		1300 Maximum growth	
Petioles	During fruit production	1105 Critical	Coltman, 1988
		1200 Optimum	
Petiole sap (adequate)	Early flower	1100–2140 (Critical value: 760)	Reuter and Robinson, 1997
	Early fruit set	1000–1200 (Critical value: 760)	
	First mature fruit	1790 (Critical value: 1120)	
	Midharvest	1600	
Petiole sap (field tomato)	First buds	1000–1200	Hartz and Hochmuth, 1996
	First open flowers	600–800	
	Fruits 2-cm diameter	400–600	
	Fruits 5-cm diameter	400–600	
	First harvest	300–400	
	Second harvest	200–400	
Petiole sap (greenhouse tomato)	Transplant to second cluster	1000–2000	
	Second to fifth cluster	800–1000	
	Harvest season (Dec–June)	700–900	
Auxillary shoots	Initial to final fruit	450–780	Morard and Kerhoas, 1984
Ammonium–Nitrogen (NH_4–N) [mg kg^{-1} (ppm)]			
Auxillary shoots	Initial to final fruit	<77	Morard and Kerhoas, 1984

Phosphorus (P)

Atomic Number: 15 **Atomic Weight:** 30.973

Discoverer of Essentiality and Year: Ville—1860

Designated Element: major element

Function: component of certain enzymes and proteins involved in energy transfer reactions andcomponent of RNA and DNA

Mobility in the Tomato Plant: mobile

Requirement Level: medium

Forms Utilized by the Tomato Plant: mono- and di-hydrogen phosphate (HPO_4^{2-} and $H_2PO_4^-$) depending on pH

Common Fertilizers:

Sources	Formula	% P (Other)
Inorganic		
Superphosphate	(0–20–0)[a]	8.74
Triple superphosphate	(0–45–0)[a]	19.66
Diammonium phosphate	$(NH_4)_2HPO_4$	21 (18 N)
Dipotassium phosphate	K_2HPO_4	18 (22 K)
Monoammonium phosphate	$(NH_4)H_2PO_4$	21 (11 N)
Monopotassium phosphate	KH_2PO_4	32 (30 K)
Phosphoric acid	H_3PO_4	34
Organic		
Bonemeal		9.6
Animal manure		0.3–0.7
Sewage sludge		1.3
Chicken litter		1.9–2.6

[a] Fertilizer designation.

Concentration in Nutrient Solution: 30–50 mg L^{-1} (ppm) of P [as either the $H_2PO_4^-$ or HPO_4^{2-} anions]; when supplied continuously, the concentration to be 5–10 mg L^{-1} (ppm) of P

Fertilizer Phosphorus Recommendation: range, 50–200 lb A^{-1} of P_2O_5 (56–224 kg ha^{-1} of P_2O_5); mean, 100 lb A^{-1} of P_2O_5 (112 kg ha^{-1} of P_2O_5)

Typical Deficiency Symptoms: slow and reduced growth, with developing purple pigmentation occurring first on the older leaves and when severe on all leaves; foliage possibly also appearing very dark green in color; low rooting medium and air temperature possibly reducing P uptake as well as creating the purple pigmentation typical of P deficiency

Symptoms of Excess: slow plant growth, with some visual symptoms frequently related to zinc (Zn) deficiency, that can be severe as large sections of the leaves will turn light brown giving the appearance of being "burned," toxicity effect possibly intensifying under anaerobic rooting conditions

Toxic Concentration in Tomato Plant: greater than 1.00% in recently mature leaves

Sufficiency Range in the Tomato Plant:

Tissue	Plant Stage	P Content (%)	Source
	Total P		
Compound leaves adjacent to top inflorescences	Midbloom	0.25–0.80	Mills and Jones, 1996
Mature leaves from new growth (greenhouse)	Mature plants	0.31–0.46	
Mature leaf	Not given	0.40–0.70	Halliday and Trenkel, 1992
Young mature leaf (optimum fertility)	Half fruit	0.50	
Whole leaf	Prior to fruiting	0.50–0.80	Hochmuth, 1996b
	During fruiting	0.40–0.60	
Youngest open leaf blade (adequate)	Early flower	0.40–0.90	Reuter and Robinson, 1997
	Early fruit set	0.30–0.70	
	First mature fruit	0.40–0.90	
	Midharvest	0.60–0.80	
	Extractable P [mg kg^{-1} (ppm)]		
Auxillary shoots	Initial to final fruit	72–109	Morard and Kerhoas, 1984

POTASSIUM (K)

Atomic Number: 19 **Atomic Weight:** 39.098

Discoverer of Essentiality and Year: von Sachs and Knop—1860

Designated Element: major element

Function: maintains ionic balance and water status in the plant; involved in the opening and closing of stomata; associated with the carbohydrate chemistry; affects fruit quality, uneven ripening and poor storage characteristics

Mobility in the Tomato Plant: mobile

Requirement Level: high

Form Utilized by the Tomato Plant: potassium (K^+) cation

Common Fertilizers:

Sources	Formula	K Other (%)
Dipotassium phosphate	K_2HPO_4	22 (18 P)
Monopotassium phosphate	KH_2PO_4	30 (32 P)
Potassium chloride	KCl	50 (47 Cl)
Potassium sulfate	K_2SO_4	42 (17 S)
Potassium nitrate	KNO_3	36 (13 N)
Sul-Po-Mag	$K_2SO_4{\cdot}2MgSO_4$	18 (11 Mg; 22.7 S)

Concentration in Nutrient Solutions: 100–200 mg L^{-1} (ppm) of K^+

Fertilizer Potassium Requirement: range, 50–300 lb A^{-1} of K_2O (56–336 kg ha^{-1} of K_2O); mean, 150 lb A^{-1} of K_2O (168 kg ha^{-1} of K_2O)

Typical Deficiency Symptoms: initially slowed growth with marginal death of older leaves giving a "burned" or "scorched" appearance; fruit yield and quality reduced; fruit postharvest quality reduced; uptake reduced under anaerobic rooting conditions and low rooting medium temperature

Symptoms of Excess: plants developing either magnesium (Mg) or calcium (Ca) deficiency symptoms as a result of a cationic imbalance

Sufficiency Range in the Tomato Plant:

Tissue	Plant Stage	K Content (%)	Source
		Total K	
Compound leaves adjacent to top inflorescences	Midbloom	2.5–5.0	Mills and Jones, 1996
Mature leaves from new growth (greenhouse)	Mature plants	3.5–5.0	
Whole leaf	Prior to fruiting	3.5–4.5	Hochmuth, 1996b
	During fruiting	2.8–4.0	
Mature leaf	Not given	2.7–5.9	Halliday and Trenkel, 1992
Young mature leaf (optimum fertility)	Half fruit	2.9	
Youngest open leaf blade (adequate)	Early flower	3.8–6.0	Reuter and Robinson, 1997
	Early fruit set	3.3–5.0	
	First mature fruit	3.0–5.0	
	Midharvest	3.4–5.2	
	Potassium in petiole sap [mg kg^{-1} (ppm)]		
Petioles (field)	First buds	3500–4000	Hartz and Hochmuth, 1996
	First open flowers	3500–4000	
	Fruits 1-in. diameter	3000–3500	

Tissue	Plant Stage	K Content (%)	Source
	Potassium in petiole sap [mg kg^{-1} (ppm)]		
	Fruits 2-in. diameter	3000–3500	
	First harvest	2500–3000	
	Second harvest	2000–2500	
Petioles (greenhouse)	Transplant to second cluster	4500–5000	
	Second to fifth cluster	4000–5000	
	Harvest season (Dec–June)	3500–4000	
Auxillary shoots	Initial to final fruit	3000-4500	Morard and Kerhoas, 1984

Calcium (Ca)

Atomic Number: 20 **Atomic Weight:** 40.07

Discoverer of Essentiality and Year: van Sachs and Knop—1860

Designated Element: major element

Function: major constituent of cell walls, and for maintaining cell wall integrity and membrane permeability; enhancing pollen germination and growth; activating a number of enzymes for cell mitosis, division, and elongation; possibly detoxifying the presence of heavy metals in tissue; affecting fruit quality, and health of conductive tissue

Mobility in the Tomato Plant: immobile

Requirement Level: high

Form Utilized by the Tomato Plant: calcium (Ca^{2+}) cation

Common Fertilizers:

Sources	Formula	Ca (Other) (%)
	Inorganic	
Gypsum	$CaSO_4 \cdot 2H_2O$	23 (19 S)
Calcium nitrate	$Ca(NO_3)_2 \cdot 4H_2O$	19 (15 N)
	Liming Materials	
Calcitic limestone	$CaCO_3$	~40
Dolomitic limestone	$Ca \cdot MgCO_3$	~22 (~13 Mg)

Concentration in Nutrient Solutions: 100–200 mg L^{-1} (ppm) of Ca^{2+}

Fertilizer Calcium Requirement: not normally specified if the soil water pH is maintained within the optimum range of 6.0–6.5

Typical Deficiency Symptoms: leaf shape and appearance changing, with the leaf margins and tips turning brown or black; vascular breakdown at the base of the plant resulting in plant wilting; occurrence of BER on fruit

Symptoms of Excess: development of either or both magnesium (Mg) and potassium (K) deficiency due to cationic imbalance

Sufficiency Range in the Tomato Plant:

Tissue	Plant Stage	Ca Content (%)	Source
		Total Ca	
Compound leaves adjacent to top inflorescences	Midbloom	1.0–3.0	Mills and Jones, 1996
Mature leaves from new growth (greenhouse)	Mature plants	1.6–3.2	
Whole leaf	Prior to fruiting	0.9–1.8	Hochmuth, 1996b
	During fruiting	1.0–2.0	
Mature leaf	Not given	2.4–7.2	Halliday and Trenkel, 1992
Young mature leaf (optimum fertility)	Half fruit	1.2	
Youngest open leaf blade (adequate)	Early flower	1.5–2.5	Reuter and Robinson, 1997
	Early fruit set	1.4–3.2	
	First mature fruit	1.4–4.0	
	Midharvest	2.0–4.3	
	Extractable Ca [mg kg^{-1} (ppm)]		
Auxillary shoots	Initial to final fruit	50–100	Morard and Kerhoas, 1984

MAGNESIUM (Mg)

Atomic Number: 12 **Atomic Weight:** 24.30

Discoverer of Essentiality and Year: von Sachs and Knop—1860

Designated Element: major element

Function: major constituent of the chlorophyll molecule; enzyme activator for a number of energy transfer reactions

Mobility in the Tomato Plant: relatively mobile

Requirement Level: medium

Form Utilized by Tomato Plant: magnesium (Mg^{2+}) cation

Common Fertilizers:

Sources	Formula	Mg (Other) (%)
Magnesium sulfate	$MgSO_4 \cdot 7H_2O$	10 (23 S)
Sul-Po-Mag	$K_2SO_4 \cdot 2MgSO_4$	11 (18 K; 22.7 S)
Dolomitic limestone	$Ca \cdot MgCO_3$	~13 (~22 Ca)

Concentration in Nutrient Solutions: 30–70 mg L^{-1} (ppm) of Mg^{2+}

Fertilizer Magnesium Requirement: if the soil water pH maintained within the optimum range of 6.0–6.5, no additional Mg normally needed; however, if soil test Mg low, then apply 25 lb A^{-1} of Mg (28 kg ha^{-1} of Mg)

Typical Deficiency Symptoms: interveinal yellowing of the older leaves; possible development of BER on fruit; uptake significantly being reduced at low (<17°C [62.7°F]) rooting temperature

Symptoms of Excess: results in cation imbalance among calcium (Ca) and potassium (K); slowed growth with possible development of either Ca or K deficiency symptoms

Sufficiency Range in the Tomato Plant:

Tissue	Plant Stage	Mg Content (%)	Source
	Total Mg		
Compound leaves adjacent to top inflorescences	Midbloom	0.4–0.9	Mills and Jones, 1996
Mature leaves from new growth (greenhouse)	Mature plants	0.4–0.5	
Whole leaf	Prior to fruiting	0.5–0.8	Hochmuth, 1996b
	During fruiting	0.4–1.0	
Mature leaf	Not given	0.4–0.9	Halliday and Trenkel, 1992
Young mature leaf (optimum fertility)	Half fruit	0.4	
Youngest open leaf blade (adequate)	Early flower	0.4–0.6	Reuter and Robinson, 1997
	Early fruit set	0.4–0.7	
	First mature fruit	0.4–1.2	
	Midharvest	0.5–1.3	
	Extractable Mg [Mg kg^{-1} (ppm)]		
Auxillary shoots	Initial to final fruit	100–150	Morard and Kerhoas, 1984

Sulfur (S)

Atomic Number: 16 **Atomic Weight:** 32.06

Discoverer of Essentiality and Year: von Sachs and Knop—1865

Designated Element: major element

Function: constituent of two amino acids, cystine and thiamin; component of compounds that give unique odor and taste to some types of plants

Form Utilized by Tomato Plant: sulfate (SO_4^{2-}) anion

Mobility in Tomato Plant: relatively mobile

Requirement Level: medium

Common Fertilizers:

Sources	Formula	S (Other) (%)
Ammonium sulfate	$(NH_4)_2SO_4$	24 (21 N)
Magnesium sulfate	$MgSO_4 \cdot 7H_2O$	23 (10 Mg)
Potassium sulfate	K_2SO_4	17 (42 K)
Gypsum	$CaSO_4 \cdot H_2O$	23 (26 Ca)
Sul-Po-Mag	$K_2SO_4 \cdot 2MgSO_4$	22.7 (18 K; 11 Mg)

Concentration in Nutrient Solution: 100 mg L^{-1} (ppm) of S [as the sulfate (SO_4^{2-}) anion]

Fertilizer Sulfur Requirement: normally S not recommended as an amendment since S is found in many commonly used fertilizers

Typical Deficiency Symptoms: general loss of green color of the entire plant; slowed growth

Symptoms of Excess: not well defined

Sufficiency Range in Tomato Plant:

Tissue	Plant Stage	S Content (%)	Source
Compound leaves adjacent to top inflorescences	Midbloom	0.3–1.2	Mills and Jones, 1996
Mature leaves from new growth (greenhouse)	Mature plants	2.8–4.2	
Whole leaf	Prior to fruiting	0.4–0.8	Hochmuth, 1996b
	During fruiting	0.4–0.8	
Mature leaf	Not given	1.0–3.2	Halliday and Trenkel, 1992
Young mature leaf (optimum fertility)	Half fruit	0.3	

THE MICRONUTRIENTS

BORON (B)

Atomic Number: 5 **Atomic Weight:** 10.81

Discoverer of Essentiality and Year: Sommer and Lipman—1926

Designated Element: micronutrient

Function in Plant: associated with carbohydrate chemistry, pollen germination, and cellular activities (division, differentiation, maturation, respiration, and growth); important in the synthesis of one of the bases for RNA formation

Mobility in Tomato Plant: immobile

Requirement Level: medium

Forms Utilized by Tomato Plant: as the borate (BO_3^{3-}) anion or nonionized molecular boric acid (H_3BO_3)

Common Fertilizers:

Sources	Formula	B (%)
Borax	$Na_2B_4O_7 \cdot 10H_2O$	11
Boric acid	H_3BO_3	16
Solubor	$Na_2B_4O_7 \cdot 4H_2O + Na_2B_{10}O_{16} \cdot 10H_2O$	20

Concentration in Nutrient Solution: 0.2–0.4 mg L^{-1} (ppm) of B [as either the borate (BO_3^{3-}) anion or nonionized boric acid (H_3BO_3)]

Fertilizer Boron Recommendation: 1 lb A^{-1} of B (1.12 kg ha^{-1} of B)

Maximum Tolerable Boron Content in Irrigation Water: 4–6 mg B L^{-1} (ppm)

Typical Deficiency Symptoms: slowed and stunted new growth, with possible death of the growing point and root tips; lack of fruit set and development

Symptoms of Excess: B accumulating in the margins of the leaf and when in excess, the margins turning black; when in sufficient excess, resulting root death

Sufficiency Range in Tomato Plant:

Tissue	Plant Stage	B Content [mg kg^{-1} (ppm)]	Source
Compound leaves adjacent to top inflorescences	Midbloom	25–75	Mills and Jones, 1996
Mature leaves from new growth (greenhouse)	Mature plants	45–76	

Tissue	Plant Stage	B Content [mg kg^{-1} (ppm)]	Source
Whole leaf	Prior to fruiting/during fruiting	35–60	Hochmuth, 1996b
Mature leaf	Not given	32–97	Halliday and Trenkel, 1992
Young mature leaf (optimum fertility)	Half fruit	25	
Youngest open leaf blade	Early flower/early fruit set	30–100	Reuter and Robinson, 1997

CHLORINE (Cl)

Atomic Number: 17 **Atomic Weight:** 35.45

Discoverer of Essentiality and Year: Broyer and others—1954

Designated Element: micronutrient

Function in Plant: involved in the evolution of oxygen (O_2) in photosystem II; raises cell osmotic pressure and affects stomatal regulation; increases hydration of plant tissue

Mobility in Tomato Plant: mobile

Form Utilized by the Tomato Plant: chloride (Cl^-) anion

Common Fertilizers:

Source	Formula	Cl (Other) (%)
Potassium chloride	KCl	47 (50 K)

Concentration in Nutrient Solution: 100 mg L^{-1} (ppm) of Cl^-

Maximum Tolerable Chloride Content in Irrigation Water: 70 mg L^{-1} (ppm) of Cl

Typical Deficiency Symptoms: chlorosis of the younger leaves; wilting

Symptoms of Excess: premature yellowing of the leaves; burning of leaf tips and margins; bronzing and abscission of leaves

Sufficiency Range in Tomato Plant: not clearly known but range for other plants from 0.5 to 2.5% for the youngest open leaf blade at early flowering

COPPER (Cu)

Atomic Number: 29 **Atomic Weight:** 64.54

Discoverer of Essentiality and Year: Lipman and Mackinnon—1931

Designated Element: micronutrient

Function in Plant: constituent of the chlorophyll protein plastocyanin; participates in electron transport system linking photosystem I and II; participates in carbohydrate metabolism and nitrogen (N_2) fixation

Mobility in Tomato Plant: immobile

Requirement Level: medium-high

Forms Utilized by the Tomato Plant: cupric (Cu^{2+}) cation

Common Fertilizers:

Source	Formula	Cu (Other) (%)
Copper sulfate	$CuSO_4 \cdot 5H_2O$	25 (13 S)

Concentration in Nutrient Solution: 0.01–0.1 mg L^{-1} (ppm) of Cu^{2+}

Fertilizer Copper Recommendation (Florida, Cu-deficient soil): 2 lb A^{-1} of Cu (2.24 kg ha^{-1} of Cu)

Typical Deficiency Symptoms: reduced or stunted growth, with distortion of the younger leaves; necrosis of the apical meristem

Symptoms of Excess: induced iron (Fe) deficiency and chlorosis; root growth ceasing and root tips turning black and dying

Sufficiency Range in Plants:

Tissue	Plant Stage	Cu Content [mg kg^{-1} (ppm)]	Source
Compound leaves adjacent to top inflorescences	Midbloom	5–20	Mills and Jones, 1996
Mature leaves from new growth (greenhouse)	Mature plants	6	
Whole leaf	Prior to fruiting/during fruiting	8–20	Hochmuth, 1996b

Tissue	Plant Stage	Cu Content [mg kg^{-1} (ppm)]	Source
Mature leaf	Not given	10–16	Halliday and Trenkel, 1992
Young mature leaf (optimum fertility)	Half fruit	7	
Youngest open leaf blade	Early flower/early fruit set	5–15	Reuter and Robinson, 1997

IRON (Fe)

Atomic Number: 26 **Atomic Weight:** 55.85

Discoverer of Essentiality and Year: von Sachs and Knop—1860

Designated Element: micronutrient

Function in Plant: component of many enzyme and electron transport systems; component of protein ferredoxin; required for nitrate (NO_3) and sulfate (SO_4) reduction, nitrogen (N_2) assimilation, and energy (NADP) production; associated with chlorophyll formation

Mobility in Tomato Plant: immobile

Requirement Level: high

Forms Utilized by Tomato Plant: ferrous (Fe^{2+}) and ferric (Fe^{3+}) cations

Common Fertilizers:

Sources	Formula	Fe (Other) (%)
Ferrous sulfate	$FeSO_4 \cdot 7H_2O$	20 (11 S)
Iron chelate	FeEDTA	6–12

Concentration in Nutrient Solution: 2–12 mg L^{-1} (ppm) of ferrous (Fe^{2+}) and ferric (Fe^{3+})

Fertilizer Iron Recommendation (Florida, Fe-deficient soils): 5 lb A^{-1} of Fe (5.6 kg ha^{-1} of Fe)

Typical Deficiency Symptoms: interveinal chlorosis of younger leaves; as deficiency intensifies, older leaves being affected and younger leaves turning yellow; deficiency developing under anaerobic conditions in the root medium

Symptoms of Excess: not known for the tomato plant

Sufficiency Range in Plants:

Tissue	Plant Stage	Fe Content [mg kg^{-1} (ppm)]	Source
Compound leaves adjacent to top inflorescences	Midbloom	40–300	Mills and Jones, 1996
Mature leaves from new growth (greenhouse)	Mature plants	84–112	
Whole leaf	Prior to fruiting/during fruiting	50–200	Hochmuth, 1996b
Mature leaf	Not given	101–291	Halliday and Trenkel, 1992
Young mature leaf (optimum fertility)	Half fruit	119	
Youngest open leaf blade	Early flower/early fruit set	60–300	Reuter and Robinson, 1997

MANGANESE (Mn)

Atomic Number: 25 **Atomic Weight:** 54.94

Discoverer of Essentiality and Year: McHargue—1922

Designated Element: micronutrient

Function in Tomato Plant: involved in the oxidation–reduction process in the photosynthetic electron transport system; photosystem II for photolysis; activates indole-3-acetic acid (IAA) oxidases

Mobility in Tomato Plant: immobile

Requirement Level: medium

Form Utilized by the Tomato Plant: manganous (Mn^{2+}) cation

Common Fertilizers:

Sources	Formula	Mn (Other) (%)
Manganese chloride	$MnCl_2 \cdot 4H_2O$	28
Manganese sulfate	$MnSO_4 \cdot H_2O$	24 (14 S)
Manganese oxide	MnO	41–68

Concentration in Nutrient Solution: 0.5–2.0 mg L^{-1} (ppm) of Mn^{2+}

Fertilizer Manganese Recommendation (Florida, Mn-deficient soils): 3 lb A^{-1} of Mn (3.36 kg ha^{-1} of Mn)

Typical Deficiency Symptoms: reduced or stunted growth, with interveinal chlorosis on younger leaves; deficiency possibly induced at low rooting temperatures

Symptoms of Excess: older leaves showing brown spots surrounded by chlorotic zone or circle; black spots appearing on stems and petioles; toxicity occurring at 1000 mg kg^{-1} (ppm) of Mn in the leaf tissue

Sufficiency Range in Plants:

Tissue	Plant Stage	Mn Content [mg kg^{-1} (ppm)]	Source
Compound leaves adjacent to top inflorescences	Midbloom	40–500	Mills and Jones, 1996
Mature leaves from new growth (greenhouse)	Mature plants	55–165	
Whole leaf	Prior to fruiting/during fruiting	50–200	Hochmuth, 1996b
Mature leaf	Not given	55–220	Halliday and Trenkel, 1992
Young mature leaf (optimum fertility)	Half fruit	76	
Youngest open leaf blade	Early flower	50–250	Reuter and Robinson, 1997
	Early fruit set	50–100	

Molybdenum (Mo)

Atomic Number: 42 **Atomic Weight:** 95.94

Discoverer of Essentiality and Year: Arnon and Stout—1939

Designated Element: micronutrient

Function in Tomato Plant: component of two enzyme systems, nitrogenase and nitrate reductase, for the conversion of nitrate (NO_3^-) to ammonium (NH_4^+)

Mobility in Tomato Plant: immobile

Requirement Level: medium

Form Utilized by Tomato Plant: molybdate (MoO_4^{2-}) anion

Common Fertilizers:

Sources	Formula	Mo (%)
Ammonium molybdate	$(NH_4)_6Mo_7O_{24}{\cdot}4H_2O$	8
Sodium molybdate	$Na_2MoO_4{\cdot}2H_2O$	39–41
Molybdenum trioxide	MoO_3	66

Concentration in Nutrient Solution: 0.05-0.2 mg L^{-1} (ppm) of MoO_4^{2-}

Fertilizer Molybdenum Recommendation (Florida, Mo-deficient soils): 0.02 lb A^{-1} of Mo (0.0224 kg ha^{-1} of Mo)

Typical Deficiency Symptoms: resembles nitrogen (N) deficiency symptoms, with older and middle leaves becoming chlorotic; rolling leaf margins; growth and flower formation restricted

Symptoms of Excess: not known

Sufficiency Range in Tomato Plant:

Tissue	Plant Stage	Mo Content [mg kg^{-1} (ppm)]	Source
Compound leaves adjacent to top inflorescences	Midbloom	>0.6	Mills and Jones, 1996
Mature leaves from new growth (greenhouse)	Mature plants	2.9–5.8	
Mature leaf	Not given	0.9–10	Halliday and Trenkel, 1992
Young mature leaf (optimum fertility)	Half fruit	0.16	
Youngest open leaf blade	Early flower	0.6	Reuter and Robinson, 1997

ZINC (Zn)

Atomic Number: 30 **Atomic Weight:** 65.39

Discoverer of Essentiality and Year: Sommer and Lipman—1926

Designated Element: micronutrient

Function in Tomato Plant: involved in same enzymatic functions as manganese (Mn) and magnesium (Mg) specific to the enzyme carbonic anhydrase

Mobility in Tomato Plant: immobile

Requirement Level: medium-high

Form Utilized by the Tomato Plant: zinc (Zn^{2+}) cation

Common Fertilizers:

Sources	Formula	Zn (Other) (%)
Zinc sulfate	$ZnSO_4 \cdot 7H_2O$	22 (11 S)
Zinc oxide	ZnO	78–80
Zinc chelates	$Na_2ZnEDTA$	14
	NaZnTA	13
	NaZnHEDTA	9

Concentration in Nutrient Solution: 0.05–0.10 mg L^{-1} (ppm) of Zn^{2+}

Fertilizer Zinc Recommendation (general range): 2–5 lb A^{-1} of Zn (2.24–5.6 kg ha^{-1} of Zn)

Typical Deficiency Symptoms: upper leaves curling with rosette appearance; chlorosis in the interveinal areas of new leaves producing a banding effect; leaves dying and falling off; flowers abscising; induced by high P levels in the rooting medium and under anaerobic rooting conditions

Symptoms of Excess: possibility of plants developing typical iron (Fe) deficiency symptoms; chlorosis of young leaves

Sufficiency Range in Tomato Plant:

Tissue	Plant Stage	Zn Content [mg kg^{-1} (ppm)]	Source
Compound leaves adjacent to top inflorescences	Midbloom	20–50	Mills and Jones, 1996
Mature leaves from new growth (greenhouse)	Mature plants	39	
Whole leaf	Prior to fruiting/during fruiting	25–60	Hochmuth, 1996b
Mature leaf	Not given	20–85	Halliday and Trenkel, 1992
Young mature leaf (optimum fertility)	Half fruit	24	
Youngest open leaf blade	Early flower/early fruit set	30–100	Reuter and Robinson, 1997

REFERENCES

Beverly, R.B. 1994. Stem sap testing as a real-time guide to tomato seedling nitrogen and potassium fertilization. *Commun. Soil Sci. Plant Anal.* 25:1045–1056.

Coltman, R.R. 1988. Yields of greenhouse tomatoes managed to maintain specific petiole sap nitrate levels. *HortScience* 23:148–151.

Halliday, D.J. and M.E. Trenkel (Eds.). 1992. *IFA World Fertilizer Use Manual*, pp. 289–290, 331–337. International Fertilizer Industry Association, Paris, France.

Hartz, T.K. and G.J. Hochmuth. 1996. Fertility management of drip-irrigated vegetables. *HortTechnology* 6(3):186–172.

Hochmuth, G.J. 1996b. Greenhouse tomato nutrition and fertilization for southern latitudes, pp. 37–39. In: *Greenhouse Tomato Seminar.* ASHS Press, American Society for Horticultural Science, Alexandria, VA.

Mills, H.A. and J. B. Jones, Jr. 1996. *Plant Nutrition Manual II.* Micro-Macro Publishing, Athens, GA.

Morard, P. and J. Kerhoas. 1984. Tomato and cucumber, pp. 677–687. In: P. Martin-Prével, J. Gagnard, and P. Gautier (Eds.), *Plant Analysis as a Guide to the Nutrient Requirements of Temperate and Tropical Crops*. Lavoisier Publishing, New York.

Reuter, D.J. and J.B. Robinson (Eds.). 1997. *Plant Analysis: An Interpretation Manual.* 2nd ed. CSIRO Publishing, Collingwood, Australia.

Appendix IV: Summary of Tomato Plant Physiological and Plant Production Characteristics and Statistics

PHYSIOLOGICAL CHARACTERISTICS

Botanical Classification:

Family—Solanaceae
Genus—*Lycopersicon* ("wolf peach") *esculentum* (tomato); *Lycopersicon pimpinellifolium* (L.) Mill (currant tomato); *Lycopersicon esculentum* var. *cerasiforme* (cherry tomato)

Historical Background: originated in the coastal strip of western South America, from the equator to about 30° latitude south, especially Peru and the Galápagos Islands, being first domesticated in Mexico, in the mid-16th century introduced into Europe

Common Early Use: featured in herbals

Early Name Designation: Moor's apple (Italian) or "love apple" (France)

Common Names: tomato (English); tomat (Danish); tomast (Dutch), tomate (French); Tomate (German); pomodoro (Italian); tomate (Portuguese); tomato (Spanish), tomat (Swedish)

Plant Form: determinate (plant terminates with fruit cluster); indeterminate (continuously producing three nodes between each inflorescence)

Plant Character: herbaceous perennial usually grown as an annual in temperate regions since it is killed by frost

Photosynthetic Characteristics: C3 plant that saturates at light levels of 13 Mj m^{-2} day^{-1}, responsive to elevated atmospheric carbon dioxide (CO_2) content up to 1000 mg L^{-1} (ppm), responsive to extended light periods a low light intensities (400–500 μmol m^{-2} sec^{-1})

Optimum Photon Flux: 20–30 mol m^{-2} day^{-1}

Net Photosynthesis Range: 56–80 μg m^{-2} sec^{-1} of CO_2

Carbon Dioxide Toxicity: >1000 mg L^{-1} (ppm) of CO_2

Degree Days: 3000–4000 (°C)

Moisture Requirements:

Field**—**2000–6000 m^3 ha^{-1}; sensitive to being waterlogged
Greenhouse**—**about 1 L/day for mature plant

Flowering Habit: day neutral, flowers in either short or long days

Flower Color: yellow

Pollination: self-pollinated

Minimum temperature**—**65°F (18.3°C)
Maximum temperature**—**85°F (29.4°C)

Days to Maturity:

Very Early**—**45–50
Early**—**50–60
Midseason**—**70–80
Late**—**85–95

FRUIT CHARACTERISTICS

Fruit Size: 2–12 locules from 2–6 in. in diameter

Fruit Color: red, pink, yellow, orange

Fruit Nutritional Content:

Lycopene—antioxidant
Vitamins—A and C
Mineral—potassium (K)

Fruit Color Classification:

Green**—**completely green
Breakers**—**definite break in color

Turning—>10%, <30% change in color
Pink—>30%, <60% pink or red
Light Red—>60%, <90% red color
Red: >90% red color

Fruit Storage Requirements:

Firm Ripe
Temperature—46–50°F (7.8–10°C)
Storage Time—1–3 weeks
Mature Green
Temperature—55–70°F (12.8–21°C)
Storage Time—4–7 weeks

Commonly Occurring Fruit Disorders: cracking, catfacing, puffiness, blossom-end rot (BER), sunscald, green shoulders, russeting, anther scarring, uneven ripening

Fruit Flavor:

Good—high acidity and high sugar
Tart—high acidity and low sugar
Bland—low acidity and high sugar
Tasteless—low acidity and low sugar

TEMPERATURE REQUIREMENTS

Optimum Air Temperature for Plant Growth:

Day—65–85°F (18.3–29.4°C)
Night—65–70°F (18.3–21°C)

Optimum Canopy Temperature: 68–73.4°F (20–23°C)

Optimum Rooting Temperature: 65–75°F (18.3–23.9°C)

Tomato Seed Characteristics:

Size—3–5 mm
Seeds per ounce—4,000–12,000

Optimum Seed Germination:

Temperature—60–85°F (15.5–29.°C)
Time—6–8 days

MINERAL NUTRITION

Major Essential Elements:

Element	Requirement Level	Normal Range in Plant (%)
	Major Elements	
Nitrogen (N)	Medium	2.8–6.0
Phosphorus (P)	Medium	0.3–0.9
Potassium(K)	High	2.5–5.0
Calcium (Ca)	High	1.0–4.0
Magnesium (Mg)	Medium	0.5–1.0
Sulfur (S)	Medium	0.3–1.2
		[mg kg^{-1} (ppm)]
	Micronutrients	
Boron (B)	Medium	25–75
Copper (Cu)	Medium-high	5–20
Iron (Fe)	High	50–200
Manganese (Mn)	Medium	50–200
Molybdenum (Mo)	Medium	0.9–5.0
Zinc (Zn)	Medium-high	25–100

Optimum Soil Water pH: 6.0–6.5

Optimum Soil Test Level (Fertility): "medium" to "high" category

Maximum Salinity: 2.5 decisemens per meter (dS m^{-1})

Fertilizer Rate Recommendation Range (Average):

Field Production Rates (per acre)

Major Elements

Nitrogen—50–220 lb N
Phosphorus—60–200 lb P_2O_5
Potassium—55–300 lb K_2O

Micronutrients

Boron—2.0 lb B
Copper—2.0 lb Cu
Iron—5.0 lb Fe
Manganese—3.0 lb Mn
Molybdenum—0.02 lb Mo
Zinc—2.0 lb Zn

Home Garden Fertilizer Rates (per 1000 ft^2):

Nitrogen—2–3 lb N
Phosphorus—2–5 lb P_2O_5
Potassium—2–5 lb K_2O

PLANT SPACING

Row Spacing in Field:

Staked—12–24 in. in row; 36–48 in. between rows
Processing—2–10 in. in row; 42–60 in. between rows
Density—3–4 m^2 per plant (12,150–36,900 plants per hectare)

Row Spacing—Home Garden:

Per 100 ft of row—35–65 plants
Between plants in row or between rows—18–36 in.

Greenhouse Plant Spacing:

Density—0.35–0.40 plants per square meter; 4 ft^2 per plant
Row Arrangement—double rows 18 in. apart, 14–16 in. between plants within the row, and 4 ft between each set of double rows

HYDROPONICS

Common Hydroponic Growing Systems:

Nutrient Film Technique (NFT)
Perlite bag drip irrigation
Rockwool slab drip irrigation

Major and Micronutrient Ionic Forms and Normal Concentration Range in the Nutrient Solution:

Element	Ionic Form	Concentration in Solution [mg L^{-1} (ppm)]
	Major Elements	
Nitrogen (N)	NO_3^- or NH_4^+	100–200
Phosphorus (P)	HPO_4^{2-} or $H_2PO_4^-$ [a]	30–50
Potassium (K)	K^+	100–200
Calcium (Ca)	Ca^{2+}	100–200
Magnesium (Mg)	Mg^{2+}	30–70

Element	Ionic Form	Concentration in Solution [mg L^{-1} (ppm)]
	Micronutrients	
Boron (B)	BO_3^{3-} or H_3BO_3 [b]	0.2–0.4
Chloride (Cl)	Cl^-	5.0
Copper (Cu)	Cu^{2+}	0.01–0.1
Iron (Fe)	Fe^{2+} or Fe^{3+}	2–12
Manganese (Mn)	Mn^{2+}	0.5–2.0
Molybdenum (Mo)	MoO_4^{2-}	0.05–0.2
Zinc (Zn)	Zn^{2+}	0.05–0.10

[a] Form dependent on pH.
[b] Increasing evidence that molecular H_3BO_3 is the form in solution.

COMMON PESTS

Common Diseases: anthracnose, bacterial canker, bacterial spot, early blight, late blight, fusarium wilt, gray leaf spot, leaf mold, mosaic, verticillium wilt

Seed-Bearing Diseases: bacterial canker, bacterial spot, bacterial speck, anthracnose

Common Insects: aphid, Colorado potato beetle, corn earworm, flea beetle, fruit fly, hornworm, leaf miner, pinworm, spider mite, stick bug, whitefly

Nematodes: Root knot

Seed Disease Resistance Identification Code:

Code	Disease
V	Verticillium wilt
F	Fusarium wilt
FF	Fusarium, races 1 and 2
N	Nematodes
T	Tobacco mosaic virus
A	Alternaria stem canker
St	Stemphylium gray leaf spot

PRODUCTION/CONSUMPTION STATISTICS

World Production (1994):

Acres (ha x 10^3)—2,852
Yield (ton ha^{-1})—27.2
Production (ton x 10^3)—77,540

World Production and Utilization:

Developing Countries

Production—38.6 × 10^6 ton

Utilization—72.7 g day^{-1}

Developed Countries

Production—33.7 × 10^6 ton

Utilization—23.9 g day^{-1}

Leading Tomato Producing States (1995):

Fresh Market—Florida, California, Georgia

Processing—California, Ohio, Indiana

Commercial United States Field Production (1995):

Fresh Market

Acres—132,820

Production—30,854 (1,000 cwt)

Processing

Acres—344,380

Production—11,282,040 ton

Fruit Yields in the United States (1995):

Processing—32.73 (ton A^{-1}), 900 cwt A^{-1}

Fresh Market—263 (cwt A^{-1}), 410 cwt A^{-1} fresh market, staked

Greenhouse—50 kg m^{-2} (88 lb yd^{-2}); 2.0 to 2.5 lb $week^{-1}$ $plant^{-1}$

Imports into United States (1995) (1,000 lb):

Fresh Market—1,702,019

Canned—221,894

Paste—33,590

Exports from United States (1995) (1,000 lb):

Fresh Market—288,021

Canned Whole—59,312

Catsup/Sauces—252,503

Paste—193,215

Juice—51,006

Per Capita United States Consumption (1995), pounds:

Fresh—15.1

Canned—79.0

Total—94.1

Index

A

B

C

D

E